ULTRASONIC SPECTROSCOPY

Ultrasonic spectroscopy is a technique widely used in solid-state physics, materials science, and geology that utilizes acoustic waves to determine fundamental physical properties of materials, such as their elasticity and mechanical energy dissipation. This book provides complete coverage of the main issues relevant to the design, analysis, and interpretation of ultrasonic experiments. Topics including elasticity, acoustic waves in solids, ultrasonic loss, and the relation of elastic constants to thermodynamic potentials are covered in depth. Modern techniques and experimental methods including resonant ultrasound spectroscopy, digital pulse-echo, and picosecond ultrasound are also introduced and reviewed. This self-contained book includes extensive background theory and is accessible to students new to the field of ultrasonic spectroscopy, as well as to graduate students and researchers in physics, engineering, materials science, and geophysics.

ROBERT G. LEISURE is Professor Emeritus in the Department of Physics at Colorado State University, where he served as chair of the Physics Department from 1984–1990. He is also Fellow of the Acoustical Society of America and the Institute of Physics. His work focuses on ultrasonic studies of solids.

ULTRASONIC SPECTROSCOPY

Applications in Condensed Matter Physics and Materials Science

ROBERT G. LEISURE
Colorado State University

CAMBRIDGE
UNIVERSITY PRESS

University Printing House, Cambridge CB2 8BS, United Kingdom

One Liberty Plaza, 20th Floor, New York, NY 10006, USA

477 Williamstown Road, Port Melbourne, VIC 3207, Australia

4843/24, 2nd Floor, Ansari Road, Daryaganj, Delhi – 110002, India

79 Anson Road, #06–04/06, Singapore 079906

Cambridge University Press is part of the University of Cambridge.

It furthers the University's mission by disseminating knowledge in the pursuit of education, learning, and research at the highest international levels of excellence.

www.cambridge.org
Information on this title: www.cambridge.org/9781107154131
DOI: 10.1017/9781316658901

First published 2017

Printed in the United Kingdom by Clays, St Ives plc

A catalogue record for this publication is available from the British Library.

Library of Congress Cataloging-in-Publication Data
Names: Leisure, Robert G., 1938– author.
Title: Ultrasonic spectroscopy : applications in condensed matter physics and materials science / Robert G. Leisure, Colorado State University.
Description: Cambridge, United Kingdom ; New York, NY : Cambridge University Press, 2017. | Includes bibliographical references and index.
Identifiers: LCCN 2016054364 | ISBN 9781107154131 (Hardback ; alk. paper) | ISBN 1107154138 (Hardback ; alk. paper)
Subjects: LCSH: Ultrasonic testing. | Materials–Testing. | Condensed matter.
Classification: LCC TA417.4 .L45 2017 | DDC 620.1/1274–dc23 LC record available at https://lccn.loc.gov/2016054364

ISBN 978-1-107-15413-1 Hardback

Contents

Preface

The use of ultrasonic methods for the study of materials continues to flourish and evolve. These methods find uses in many areas including fundamental condensed matter physics, materials science, various branches of engineering, geophysics, and applied studies of device-related material parameters. Advancements in experimental methods, especially resonant ultrasound spectroscopy, have enabled quantitative measurements on dramatically reduced specimen sizes, thereby vastly expanding the possibilities for the study of novel materials. The title of the book "*Ultrasonic Spectroscopy*" is taken here to mean simply the investigation of material properties by the use of ultrasonic waves.

A major purpose of this book is to present an in-depth coverage of the main issues underlying the planning and interpretation of ultrasonic investigations of materials. It is intended that the level of the presentation be accessible to dedicated upper-level undergraduate students, but at the same time achieve a depth of coverage useful to graduate students and other researchers. The approach is to present in careful detail a number of topics, with two objectives in mind. One objective, of course, is to educate the reader about basic concepts in the field – concepts that should become familiar to any researcher in this area. A second objective, perhaps more important, is to illustrate theoretical ideas that can be applied to a wide variety of problems. The emphasis is on basic concepts, not specific materials. The goal is to provide a fundamental background for beginning researchers so that – with the help of a good scientific Internet search engine to obtain more focused information – they are able to attack any of the gamut of interesting problems amenable to ultrasonic methods.

The mathematical methods used should be familiar to upper-level undergraduate students. Some knowledge of thermodynamics, statistical mechanics, and solid-state physics is required, but an effort is made to present the key concepts from these subjects as needed, and provide references to more detailed sources.

The book includes one chapter on experimental methods. Both continuous wave and pulse techniques are discussed.

I have benefited from interactions with many people over the years, too many to list here. Special thanks go to Albert Migliori, Hassel Ledbetter, Ricardo Schwarz, Paul Heyliger, David Hsu, Tatsuo Kanashiro, Alan Levelut, Jean-Yves Prieur, Ori Yeheskel, many wonderful students, and my thesis advisor, the late Dan Bolef. Special thanks also go to Dennis Agosta, Kate Ross, and Frank Willis for reading large parts of the manuscript.

Most of all, I am especially thankful to my wife, Jeanine Smith Leisure, for her contributions through steadfast support, encouragement, and thought-provoking discussions.

1
Introduction

The present chapter will serve as an overview of the material to be presented in the rest of the book. While it is hoped that the material will prove useful to all those involved in or interested in the use of ultrasound as a probe of condensed matter, a special effort is made to present the material in sufficient detail so as to be helpful to dedicated, upper-level undergraduate students and beginning graduate students. Scientists from several different disciplines are nowadays finding ultrasonic spectroscopy a useful tool, thus a strong background in solid-state physics, statistical physics, and quantum mechanics is not assumed of the readers. Brief background material is presented as needed. Several monographs have contributed to the advancement of ultrasonic spectroscopy, among them References [1, 2, 3]. The author is deeply indebted to those who have helped develop the field of ultrasonic studies of materials.

Chapter 2 deals with classical elasticity; the solid is treated as a continuum. The continuum approximation is valid for virtually all ultrasonic experiments. The present treatment of elasticity is more extensive than is usually found in books on ultrasonic techniques, but this more extensive treatment seems important if the researcher is to understand the widest implications of her/his ultrasonic research. Basic physical parameters in this chapter are stress (a two-index tensor), strain (a two-index tensor), and elastic constants (a four-index tensor, which by Hooke's Law connects stress and strain). Thus, many indices and sums over these indices appear frequently. For pedagogical reasons, it was decided *not* to use the elegant Einstein summation convention. For those new to the field, it seems better to write out the sums explicitly. The relation of elastic constants to thermodynamic potentials is derived. The condensed (Voigt) notation for stress, strain, and elastic constants is explained in detail. Coordinate transformations are treated. The form of the elastic constant matrix for each of the seven crystal systems is derived, as well as the form for the icosahedral quasicrystal. Poisson's ratio and the practical moduli – bulk, Young's, and torsion – are discussed for various crystal symmetries. It is difficult to visualize the directional dependence of elastic constants for anisotropic materials; so, 3D representational surfaces are presented for several crystal symmetries.

Chapter 3 treats acoustic waves in solids. The first part of the chapter deals with waves in the continuum limit, the regime of ultrasonic experiments. The wave equation is derived and the Christoffel equation for plane wave propagation is found. It is shown how to calculate the wave velocities for any direction in a crystal. Solutions are given for several crystalline directions in cubic and hexagonal symmetries. Other symmetries are discussed. The Christoffel equation is valid in the continuum limit and thus is restricted to wave vectors near the center of the Brillouin zone. The second part of Chapter 3 treats lattice dynamics. To understand many properties of solids it is necessary to go beyond the continuum limit; lattice dynamics does just that. The usual 1D models for monatomic and diatomic cases are treated and the dispersion relations obtained. Much insight is gained from these models. However, it seems important to go beyond the 1D case, thus the full 3D lattice dynamics model is treated. The relevant equations are derived and force constant matrices are obtained for the face-centered cubic symmetry. A full 3D calculation is performed, demonstrating the dispersion relations in various high-symmetry directions. The last part of Chapter 3 discusses the highly simplified, but highly successful, Debye model of solids. The thermal energy and the specific heat are obtained, and it is shown how to calculate the Debye temperature from the elastic constants.

Chapter 4 deals with common experimental methods for measuring ultrasonic attenuation (or internal friction) and elastic constants (or ultrasound velocities). The material in Chapters 2 and 3 was needed before the experimental methods could be treated meaningfully. The well-known pulse-echo method is discussed. For highly accurate results it is necessary to accurately determine the time delay between echoes. This is not a trivial problem, and methods (pulse superposition, pulse-echo overlap) have been developed to solve this problem. However, several factors other than round-trip travel time usually contribute to the measured delay time. These factors are discussed in detail as well as how to account for them. The older methods were analog in nature. An important recent advance has been the development of an all digital pulse-echo system. Resonant ultrasound spectroscopy (RUS) is a resonance method dramatically different from the pulse-echo method. RUS is also discussed at length in Chapter 4. Finally, picosecond ultrasound is discussed in Chapter 4.

Chapter 5 treats the important subject of elastic constants. It starts with a review of relevant thermodynamics and statistical mechanics, culminating in the Helmholtz free energy for a harmonic oscillator. The individual lattice vibration modes are treated as harmonic oscillators, thus the Helmholtz free energy for the lattice is a sum of harmonic oscillators. A major part of Chapter 5 is concerned with using the quasiharmonic approximation to find the temperature dependence of the elastic constants. This rather lengthy calculation reveals the indirect effect of lattice expansion on the temperature dependence of the elastic constants and shows that it is of major importance. The effects of both first-order and second-order phase transitions on the elastic constants are discussed in terms of the Landau theory.

Chapter 6 is concerned with ultrasonic loss. The chapter begins with several general ideas related to ultrasonic loss: complex elastic constants and phase shifts; units of measurement of loss; Kramers-Kronig relations; and fluctuations and dissipation. Next, relaxational and resonance attenuation are discussed in general terms. Many different specific loss mechanisms are known. In general, these have been treated elsewhere. The present approach is to give a brief summary of ten of these loss mechanisms, with references to the literature and previous monographs for more detailed descriptions.

2
Elasticity

Chapter 2 deals with the elastic response of materials, where elastic has the usual meaning that a material returns to its original configuration after the removal of a deformation-causing force. In fact, *linear elasticity* will be assumed, for which the distortion from the equilibrium configuration is directly proportional to the applied force. In the present chapter, classical elasticity will be discussed, in which case the solid is treated as a continuum, and the discreteness of the underlying lattice is ignored [4]. This approach is only valid when the length scale of the spatial variations is much greater than the lattice constant, the usual situation for ultrasonic vibrations in solids. Knowledge of classical elasticity is essential for an understanding of ultrasonic vibrations and the propagation of ultrasonic waves in materials.

2.1 Strain

2.1.1 The Strain Tensor

The problem at hand is to describe the distortions of a material from the equilibrium configuration. Consider a point in a solid relative to an origin fixed in space. Before deformation, the position of the point is given by $\mathbf{r}$. After deformation, the position is given by $\mathbf{r}' = \mathbf{r} + \mathbf{u}$. Thus, the displacement of a point from its equilibrium position is

$$\mathbf{u} = \mathbf{r}' - \mathbf{r}. \tag{2.1}$$

In the case of interest, $\mathbf{u}$ is a function of $\mathbf{r}$, *i.e* $\mathbf{u}(\mathbf{r})$, otherwise the displacement just corresponds to a uniform displacement of the material, and is not relevant to the present discussion. A brief discussion of $\mathbf{u}(\mathbf{r})$ follows, leading to an expression for the strain tensor.

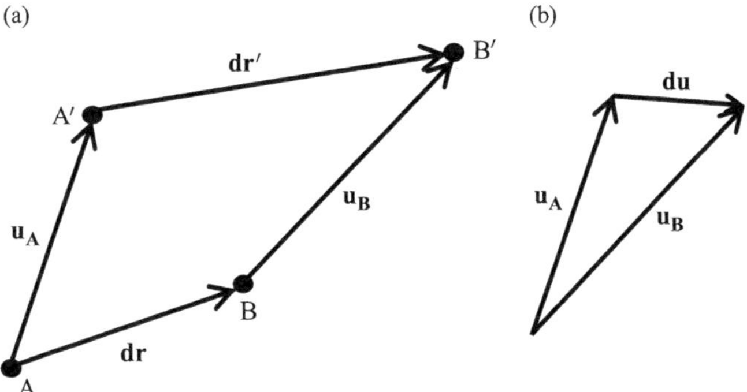

Figure 2.1 (a) The two points A and B are initially separated by the vector **dr**. After a distortion the points are separated by the vector **dr**′. (b) The vector **du** is shown.

Figure 2.1 shows two points in the material, A and B, which are originally separated by **dr**. After a distortion the points move to A′ and B′ respectively and are separated by **dr**′. The various vectors are related by

$$\mathbf{dr}' = \mathbf{dr} + (\mathbf{u_B} - \mathbf{u_A}) = \mathbf{dr} + \mathbf{du} \tag{2.2}$$

where **du** is the *difference* in the displacement vectors for the two original points, A and B. An expression for the strain is obtained by considering the difference in the square of the distances between the two points, and how it changes with the distortion [5], [6]. This quantity will be unchanged for pure rotations, but will change for distortions of the material. Squaring Equation 2.2 gives

$$(\mathbf{dr}')^2 = (\mathbf{dr})^2 + 2\mathbf{dr} \cdot \mathbf{du} + (\mathbf{du})^2. \tag{2.3}$$

It is convenient to express the equations in component form, *e.g.* $\mathbf{dr} \cdot \mathbf{du} = \sum_{i=1}^{3} dx_i du_i$. In general du_i will be a function of **r**, leading to

$$du_i = \sum_{j=1}^{3} \left(\frac{\partial u_i}{\partial x_j} dx_j \right) \tag{2.4}$$

with $i = 1, 2, 3$. Then, Equation 2.3 becomes

$$(\mathbf{dr}')^2 = (\mathbf{dr})^2 + 2 \sum_{ij} \left(\frac{\partial u_i}{\partial x_j} dx_j \right) dx_i + \sum_{ijl} \left(\frac{\partial u_i}{\partial x_j} dx_j \right) \left(\frac{\partial u_i}{\partial x_l} dx_l \right) \tag{2.5}$$

where the sums over all indices range independently from 1 to 3. Rearranging the summation indices yields the desired result. First, the second term of the right-hand side of Equation 2.5 may be rewritten as

$$2\sum_{ij}\left(\frac{\partial u_i}{\partial x_j}dx_j\right)dx_i = \sum_{ij}\left(\frac{\partial u_i}{\partial x_j}dx_j\right)dx_i + \sum_{ij}\left(\frac{\partial u_j}{\partial x_i}dx_i\right)dx_j \tag{2.6}$$

where dummy summation labels have been interchanged in the last term of Equation 2.6. Similarly, i and l can be interchanged in the last term of Equation 2.5. Finally, the result is

$$(\mathbf{dr}')^2 - (\mathbf{dr})^2 = \sum_{ij} 2\epsilon_{ij}dx_i dx_j \tag{2.7}$$

where

$$\epsilon_{ij} = \frac{1}{2}\left(\frac{\partial u_i}{\partial x_j} + \frac{\partial u_j}{\partial x_i} + \sum_l \frac{\partial u_l}{\partial x_j}\frac{\partial u_l}{\partial x_i}\right) \tag{2.8}$$

and ϵ_{ij} is the Lagrangian [7] strain tensor. For small deformations the last term in Equation 2.8 is neglected with the result that the elastic strain tensor is given by

$$\epsilon_{ij} = \frac{1}{2}\left(\frac{\partial u_i}{\partial x_j} + \frac{\partial u_j}{\partial x_i}\right). \tag{2.9}$$

Notice that the strain tensor is symmetric, $\epsilon_{ij} = \epsilon_{ji}$.

Equation 2.9 is the basic definition of infinitesimal strain that will be used throughout this work. In a few cases it will be necessary to use Equation 2.8 instead; and, Equation 2.8 will be referenced at those points.

Actually, Equation 2.9 could be taken as the starting point for the concept of strain. In what follows, the concept will be developed further.

2.1.2 Geometrical Interpretation

One-Dimensional Example

A few simple examples will be given to illustrate the meaning of the strain tensor. First, a one-dimensional distortion is discussed.

As shown in Figure 2.2, the points A and B are originally separated by a distance Δx_1. After a stretch along the x_1-axis, the two points, A′ and B′, are separated by $\Delta x_1 + \Delta u_1$. The strain in this 1D example is given by

$$\epsilon_{11} = \frac{1}{2}\left(\frac{\partial u_1}{\partial x_1} + \frac{\partial u_1}{\partial x_1}\right) = \frac{\partial u_1}{\partial x_1} \simeq \frac{\Delta u}{\Delta x}. \tag{2.10}$$

Thus, ϵ_{11} gives the fraction change in the length separating two nearby points in the $\hat{\mathbf{x}}_1$ direction (where, as usual $\hat{\mathbf{x}}_i$ denotes a unit vector lying along the x_i axis).

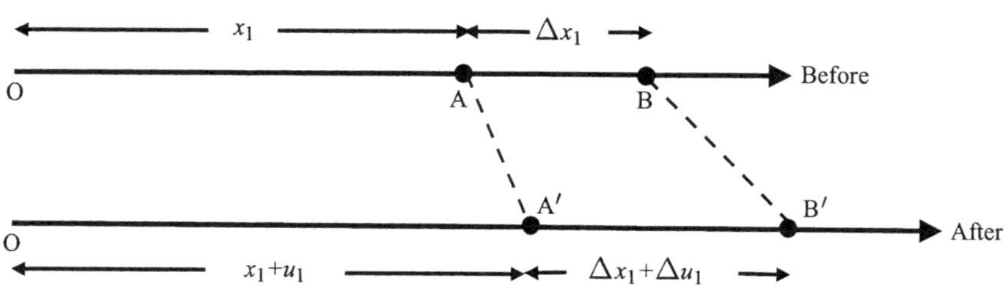

Figure 2.2 The two points A and B are initially separated by Δx_1. After a distortion the points are separated by $\Delta x_1 + \Delta u_1$.

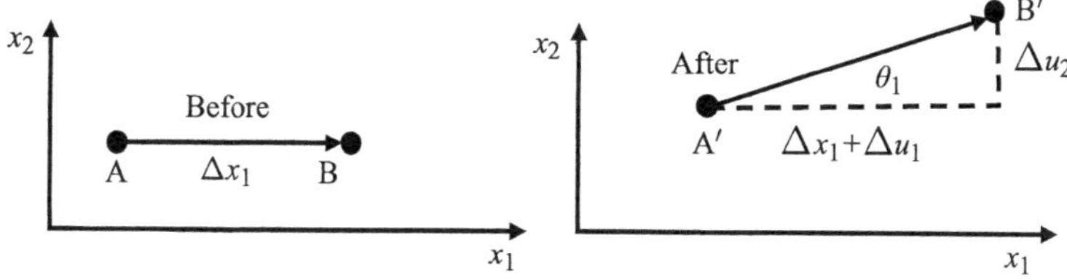

Figure 2.3 A distortion in two dimensions carries the points A and B to A′ and B′ respectively.

Two-Dimensional Examples

The preceding discussion will now be extended to two dimensions. Figure 2.3 illustrates the situation. Points A and B are carried to A′ and B′ respectively by a distortion.

The angle θ_1 is exaggerated for clarity. For small strains, θ_1 is correspondingly small. Proceeding as for Equation 2.10 shows that $\epsilon_{11} = \frac{1}{2}\left(\frac{\partial u_1}{\partial x_1} + \frac{\partial u_1}{\partial x_1}\right) = \frac{\partial u_1}{\partial x_1} \simeq \frac{\Delta u_1}{\Delta x_1}$, an extension per unit length in the x_1 direction. Inspection of Figure 2.3 reveals that

$$\frac{\partial u_2}{\partial x_1} \simeq \frac{\Delta u_2}{\Delta x_1} \simeq \theta_1. \tag{2.11}$$

Thus, $\partial u_2/\partial x_1$ represents a counterclockwise rotation by an angle θ_1 of a line originally lying parallel to the x_1 axis. A similar consideration of a line originally lying parallel to the x_2 axis shows that

$$\frac{\partial u_1}{\partial x_2} \simeq \frac{\Delta u_1}{\Delta x_2} \simeq \theta_2, \tag{2.12}$$

represents a clockwise rotation of the line.

Still discussing 2D examples, if the angle between two lines is $\pi/2$ before a homogeneous deformation, afterward it is

$$\psi = \frac{\pi}{2} - (\theta_1 + \theta_2) = \frac{\pi}{2} - \left(\frac{\partial u_2}{\partial x_1} + \frac{\partial u_1}{\partial x_2}\right) = \frac{\pi}{2} - 2\epsilon_{12} \tag{2.13}$$

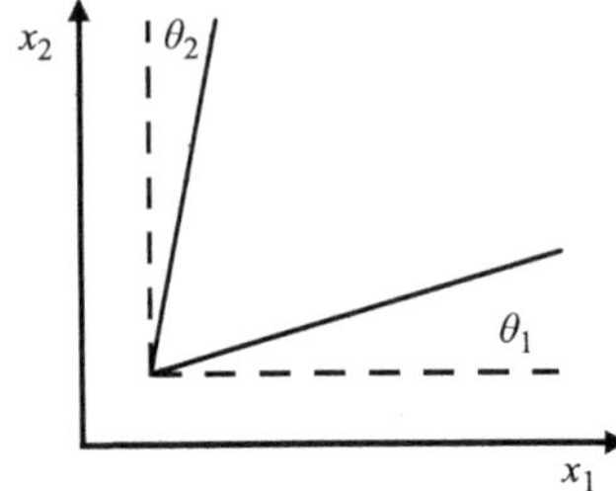

Figure 2.4 The dashed lines indicate two lines that are perpendicular before a distortion. After the distortion, the lines are rotated as shown.

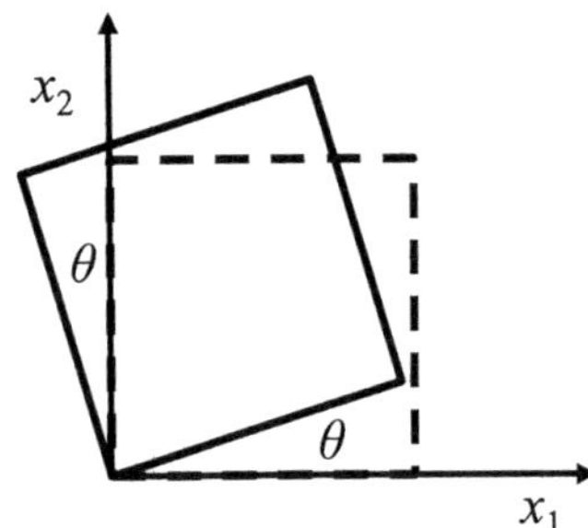

Figure 2.5 The dashed lines indicate the initial position of a square sheet of the material. The sides are parallel to the coordinate axes and one corner is located at the origin. After a rigid-body rotation about the origin of an angle θ, the sheet is oriented as shown.

as illustrated if Figure 2.4. Thus, $2\epsilon_{12}$ gives the change in the angle between two lines that were perpendicular before the deformation. The illustration in Figure 2.4 is for $\partial u_1/\partial x_2$ and $\partial u_2/\partial x_1$ both > 0. Note that $\partial u_2/\partial x_1$ need not be equal to $\partial u_1/\partial x_2$.

Rotations

Consider the situation in which a sheet of material undergoes a rigid-body rotation. The situation is illustrated in Figure 2.5. By the arguments leading to Equations 2.11 and 2.12 the displacement of the material from the original position is described by $\partial u_2/\partial x_1 = \theta$ and $\partial u_1/\partial x_2 = -\theta$. (The small-angle approximation is assumed.) From these results it follows that the strain, as defined by Equation 2.9, is zero. However, consider the quantity ω_{ij} defined as

$$\omega_{ij} = \frac{1}{2}\left(\frac{\partial u_i}{\partial x_j} - \frac{\partial u_j}{\partial x_i}\right) \tag{2.14}$$

Applying Equation 2.14 to the present situation gives $\omega_{12} = -\theta$. Thus, ω_{12} represents a rotation; counterclockwise (clockwise) for ω_{12} negative (positive). It follows

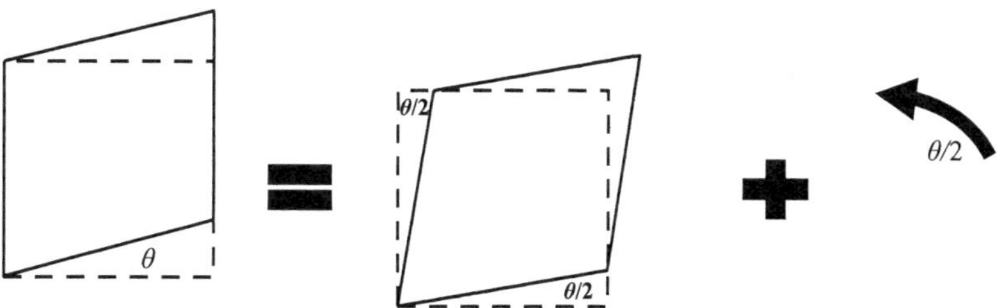

Figure 2.6 The dashed lines indicate the initial position of a square sheet of the undeformed material. A simple shear deformation, characterized by the angle θ, may be regarded as a pure shear strain plus a rotation. (The angles are exaggerated for clarity.)

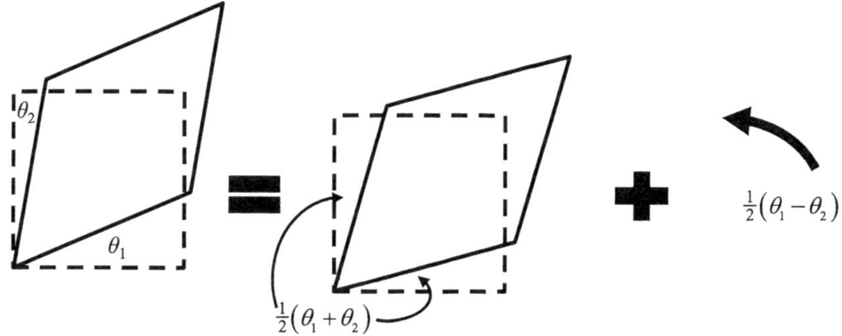

Figure 2.7 A general shear deformation, characterized by the angles θ_1 and θ_2, also may be regarded as a pure shear strain plus a rotation about an axis perpendicular to the plane.

that a general distortion of the form $\partial u_i/\partial x_j$ can be written as the sum of a symmetric and an antisymmetric part,

$$\frac{\partial u_i}{\partial x_j} = \frac{1}{2}\left(\frac{\partial u_i}{\partial x_j} + \frac{\partial u_j}{\partial x_i}\right) + \frac{1}{2}\left(\frac{\partial u_i}{\partial x_j} - \frac{\partial u_j}{\partial x_i}\right) \tag{2.15}$$

The first term on the right-hand sign of Equation 2.15 gives the tensor strain and the second term represents the rotation. Figure 2.6 illustrates the relationship for the case of $\partial u_2/\partial x_1 = \theta$ and $\partial u_1/\partial x_2 = 0$. Figure 2.6 shows how a simple shear can be decomposed into a pure shear and a rotation. A transverse ultrasonic wave propagating along a high-symmetry direction produces such a simple shear motion. The rotation is commonly ignored, because in most cases there is no energy cost for the pure rotation. However, such is not the case if an external torque acts on the material (*e.g.* a magnetic material in a magnetic field [8]). In such cases, the rotational term may become important.

Figure 2.7 illustrates a more general situation. For this case the general shear distortion on the left side of Figure 2.7, $\partial u_2/\partial x_1 = \theta_1$ and $\partial u_1/\partial x_2 = \theta_2$, can be represented as a pure shear plus a rotation as shown.

2.1.3 Discussion of Strain in Three Dimensions

It is straightforward to generalize the interpretations of Section 2.1.2 to 3D. If **u** represents the displacement of a point in the material from its equilibrium position (relative to a coordinate system fixed in space), then the distortion of the material is given by a strain ϵ_{ij}, Equation 2.9, and a rotation ω_{ij}, Equation 2.14. These have the following interpretations [4]:

ϵ_{ii} is the extension per unit length of a line segment parallel to the x_i axis,

ϵ_{ij}, $i \neq j$ equals 1/2 the change in the angle between two line elements originally parallel to axes x_i and x_j respectively. (ϵ_{ij} is the average of the angles of rotation for the two line elements). The angle decreases for $\epsilon_{ij} > 0$.

ω_{ij} represents a rotation of the material about an axis along x_k. The i, j, and k directions are assumed to represent a right-handed Cartesian coordinate system, as usual.

The Strain Tensor

For small deformations it is possible to express the local displacements from equilibrium as

$$\Delta u_i = \sum_j \frac{\partial u_i}{\partial x_j} \Delta x_j. \tag{2.16}$$

For the case of no rotations ($\partial u_i/\partial x_j = \partial u_j/\partial x_i$), application of Equations 2.9 and 2.16 give the local displacements from equilibrium as

$$\begin{aligned} \Delta u_1 &= \epsilon_{11}\Delta x_1 + \epsilon_{12}\Delta x_2 + \epsilon_{13}\Delta x_3 \\ \Delta u_2 &= \epsilon_{21}\Delta x_1 + \epsilon_{22}\Delta x_2 + \epsilon_{23}\Delta x_3 \\ \Delta u_3 &= \epsilon_{31}\Delta x_1 + \epsilon_{32}\Delta x_3 + \epsilon_{33}\Delta x_3. \end{aligned} \tag{2.17}$$

where $\Delta \mathbf{u}$ is the displacement from equilibrium at the position $\Delta \mathbf{x}$. Equation 2.17 can also be written as

$$\begin{pmatrix} \Delta u_1 \\ \Delta u_2 \\ \Delta u_3 \end{pmatrix} = \begin{pmatrix} \epsilon_{11} & \epsilon_{12} & \epsilon_{13} \\ \epsilon_{21} & \epsilon_{22} & \epsilon_{23} \\ \epsilon_{31} & \epsilon_{32} & \epsilon_{33} \end{pmatrix} \begin{pmatrix} \Delta x_1 \\ \Delta x_2 \\ \Delta x_3 \end{pmatrix}. \tag{2.18}$$

Equation 2.18 gives a linear relation between the vectors Δu_i and Δx_j, *which shows that* ϵ_{ij} *is a tensor of the second rank*, [4] the strain tensor. It is a symmetric tensor because ϵ_{ij} is symmetric by its definition (Equation 2.9). In general Δx is very small (infinitesimal), but for homogeneous strains such is not *necessarily* the case.

Simple Examples of Pure Strain Deformations (No Rotations)

Equations 2.17 will now be used to illustrate the distortion produced by pure strains. For simplicity in illustrations, 2D examples will be used. Consider the case with

ϵ_{11} being the only strain. Equations 2.17 (or Equations 2.18) give the resulting displacements, and the situation is illustrated in Figure 2.8b, *i.e.* a simple extension. Next, consider the case of $\epsilon_{12} = \epsilon_{21}$ being the only strain. If Equations 2.17 are applied to each corner of the square indicated by the dashed lines in Figure 2.8c, the result is the distortion illustrated by the solid line, a pure shear distortion.

Such considerations of distortions produced by different strains can be useful in the interpretation of ultrasonic measurements. For example, in a cubic crystal the [100]-type directions are equivalent. In general, applied strains – such as those produced by ultrasonic waves – lower the crystalline symmetry. An ϵ_{11} strain renders the [100]-type directions inequivalent, while a ϵ_{12} strain does not. With the use of Equations 2.17, it is straightforward to determine the distortions produced by other strains, *e.g.* $\epsilon_{22} - \epsilon_{11}$, a strain produced by certain ultrasonic waves. The ultrasonic waves often consist of both a pure strain and a rotation, but as mentioned earlier, the rotational part may be safely ignored in many cases.

Coordinate Transformations

Figure 2.8 can be understood from a different point of view. Because the strain, ϵ, is a second-rank tensor, it obeys the usual rule for coordinate transformations of such tensors. Let x_i' represent a new coordinate system rotated with respect to the x_i system. The transformation of the strain tensor from the x_i coordinate system to the x_i' system is given by [4]

$$\epsilon_{ij}' = \sum_{kl} a_{ik} a_{jl} \epsilon_{kl} \tag{2.19}$$

where a_{ik} is a direction cosine; the first subscript refers to the prime system while the second subscript refers to the original system. For example, a_{21} is the cosine of the angle between x_2' and x_1.

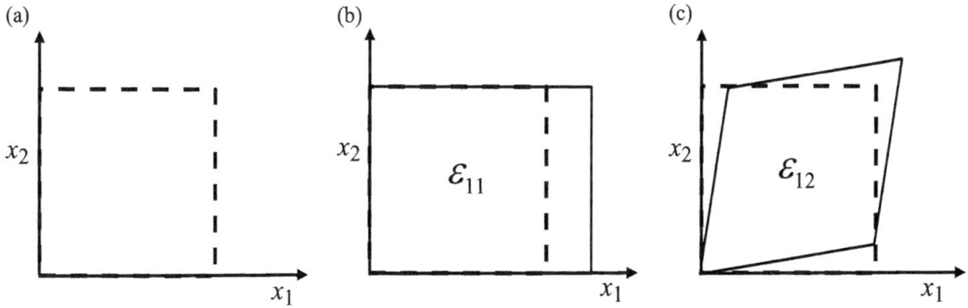

Figure 2.8 The dashed lines indicate the initial position of a square sheet of the undeformed material. The solid lines show the shapes after deformation. (a) Initial shape. (b) Deformed shape for a simple extension, ϵ_{11}. (c) Deformed shape for a pure shear, ϵ_{12}.

Now consider the case of Figure 2.8c for which the strain matrix may be written

$$\epsilon = \begin{pmatrix} 0 & \epsilon_o & 0 \\ \epsilon_o & 0 & 0 \\ 0 & 0 & 0 \end{pmatrix} \tag{2.20}$$

in the unprimed system, where

$$\epsilon_o = \frac{1}{2}\left(\frac{\partial u_1}{\partial x_2} + \frac{\partial u_2}{\partial x_1}\right) \tag{2.21}$$

Next, a counterclockwise rotation about the x_3 axis by $\pi/4$ is carried out by applying Equation 2.19 to the matrix of Equation 2.20. The result is

$$\epsilon' = \begin{pmatrix} \epsilon_o & 0 & 0 \\ 0 & -\epsilon_o & 0 \\ 0 & 0 & 0 \end{pmatrix} \tag{2.22}$$

for the strain in the primed system (coordinate axes not shown). The interpretation is simple; a pure shear in the $x_1 - x_2$ plane in the x_i system becomes an extension along x_1' axis and a compression along the x_2' axis in the x_i' coordinate system. An examination of Figure 2.8c shows that this is exactly what happens.

Principal Axes

The preceding discussion is a special case of a general result. The general symmetrical matrix of Equation 2.18 can be diagonalized to find the eigenvalues and the principal axes [9]. In the principal axes system the matrix of Equation 2.18, ϵ_{ij}, becomes

$$\epsilon = \begin{pmatrix} \epsilon_\alpha & 0 & 0 \\ 0 & \epsilon_\beta & 0 \\ 0 & 0 & \epsilon_\gamma \end{pmatrix} \tag{2.23}$$

where α, β and γ refer to the principal axes. Thus, there are no shear strains in the principal axes system, only extensions or contractions along a principal axes. As a simple example, application of the diagonalization method to the matrix of Equation 2.20 shows the principal axes to be rotated counterclockwise by $\pi/4$ about the x_3 axes with the eigenvalues of Equation 2.22, the same result achieved in the previous section by a different method. If the strain is homogeneous, then the principal axes will be the same throughout the material; otherwise, the principal axes may vary with position in the specimen.

Engineering Strains

One sometimes encounters "engineering strains," γ_{ij} which are defined somewhat differently from the tensor strains of Equation 2.9, the difference being that the 1/2 is omitted for $i \neq j$. The result is

$$\begin{aligned} \gamma_{ii} &= \frac{\partial u_i}{\partial x_j} = \epsilon_{ii} \\ \gamma_{ij} &= \frac{\partial u_i}{\partial x_j} + \frac{\partial u_j}{\partial x_i} = 2\epsilon_{ij} \quad i \neq j. \end{aligned} \tag{2.24}$$

A cautionary note, an array formed from γ_{ij} does not form a tensor, unlike the ϵ_{ij} [4]. The consequence is that ϵ_{ij} follows the tensor rules for transformation between coordinate systems, whereas γ_{ij} does not.

2.2 Stress

2.2.1 Definition of Stress

The deformations of a material are brought about by forces. Considering a particular small volume of the material, the forces acting on this volume are assumed to arise from the material outside the volume, and to act at the surface of the selected volume, *i.e.* the molecular forces are assumed to be of short range [5]. Long-range forces, (body forces) such as gravity which act throughout the material are neglected in the present treatment. It is convenient to consider forces per unit area, which are called stresses. The situation is illustrated in Figure 2.9 for the situation of a small cube which is imagined to be embedded in the material [4].

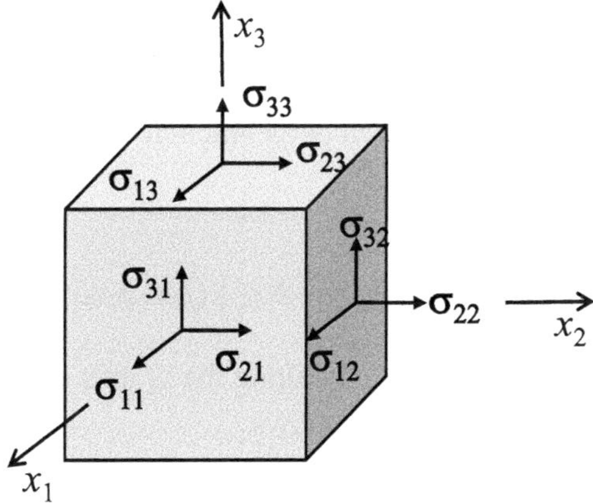

Figure 2.9 The stresses acting on a small cube of material are illustrated. σ_{ij} is the force per unit area acting in the direction of the x_i axis on a face perpendicular to the x_j direction.

In general, the force acting on a particular face of the cube will not be normal to that face, hence there will be three components of stress acting on each face. The stress, σ_{ij}, is defined as the force per unit area acting in the direction of the x_i axis on a face perpendicular to the x_j direction. For example, σ_{23} is the force per unit area acting in the x_2 direction on a face which is perpendicular to the x_3 direction. The stresses σ_{11}, σ_{22}, and σ_{33} are normal components of the stress, the others are shear stresses. The stresses are exerted on the cube by the surrounding material. By convention, positive normal components are taken as pointing outward. The positive directions of the shear stresses then follow as shown in Figure 2.9.

2.2.2 Homogeneous Stress and Static Equilibrium

In the case that the stress is homogeneous throughout the body, the stresses on opposite faces (not shown in Figure 2.9) are equal in magnitude and opposite in direction, *i.e.* the net force acting on the cube is zero. Further, if static equilibrium is required, then the net torque must be zero. An examination of Figure 2.9 shows, *e.g.* that for the torque about the x_3 axis to be zero, σ_{12} must equal σ_{21}. More generally,

$$\sigma_{ij} = \sigma_{ji}. \tag{2.25}$$

2.2.3 Inhomogeneous Stress and Equations of Motion

In general, the stress may vary throughout the material; in fact, this situation occurs for many of the most interesting cases. Consideration of the stress variation across a small cube of material leads to Newton's Second Law for linear and rotational motion of the cube. Referring to Figure 2.10 and considering first the forces in the x_1 direction, it is seen that the variation of σ_{11} leads to a force

$$\Delta F_{x_1 11} = (\Delta\sigma_{11})(\text{Area perpendicular to the } x_1 \text{ axis}),$$

or

$$\Delta F_{x_1 11} = \left(\frac{\partial \sigma_{11}}{\partial x_1}\Delta x_1\right)(\Delta x_2 \Delta x_3). \tag{2.26}$$

In a similar manner, the variation of σ_{12} along the x_2 axis leads to a net force in the x_1 direction. Reference to Figure 2.11 shows

$$\Delta F_{x_1 12} = \left(\frac{\partial \sigma_{12}}{\partial x_2}\Delta x_2\right)(\Delta x_1 \Delta x_3). \tag{2.27}$$

Of course there is a similar equation for the variation of σ_{13} along the x_3 direction.

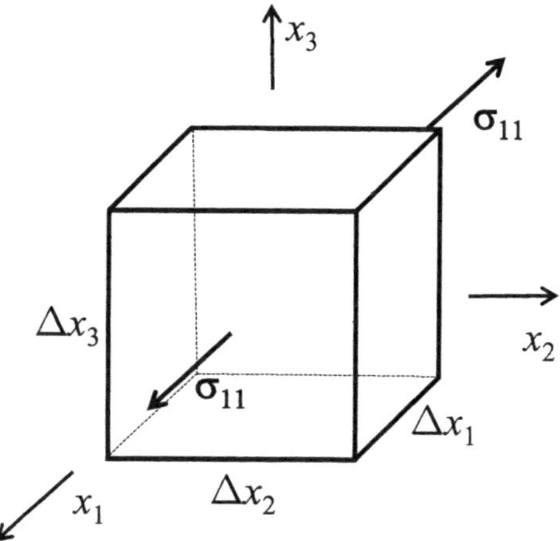

Figure 2.10 The stress σ_{11} is assumed to vary in the x_1 direction. The force, due to σ_{11}, acting on a face perpendicular to the x_1 axis is $\sigma_{11}\Delta x_2 \Delta x_3$. Variation of this force across the cube leads to a net force in the x_1 direction.

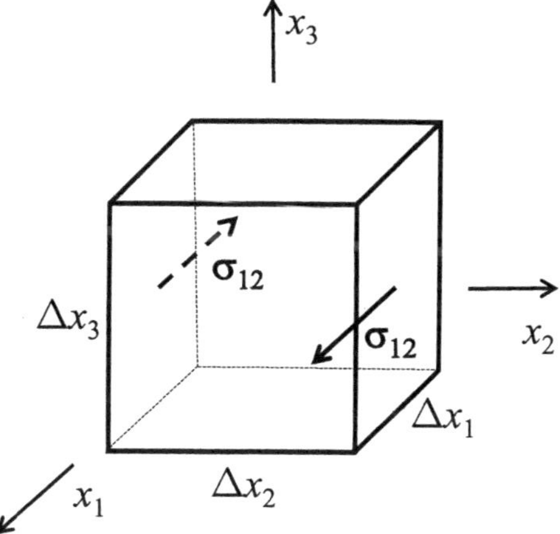

Figure 2.11 The stress, σ_{12} is assumed to vary in the x_2 direction. The force, due to σ_{12}, acting on a face perpendicular to the x_2 axis is $\sigma_{12}\Delta x_1 \Delta x_3$. Variation of this force across the cube leads to a net force in the x_1 direction.

Totaling the three contributions to the force in the x_1 direction, the equation of motion, $F_{x_1} = ma_{x_1}$ becomes

$$\left(\frac{\partial \sigma_{11}}{x_1}\right) + \left(\frac{\partial \sigma_{12}}{x_2}\right) + \left(\frac{\partial \sigma_{13}}{x_3}\right) = \left(\rho \frac{\partial^2 u_1}{\partial t^2}\right) \tag{2.28}$$

where $m = \rho(\Delta x_1 \Delta x_2 \Delta x_3)$, with ρ being the density. Also, $a_{x_1} = \partial^2 u_1/\partial t^2$. A common factor of $(\Delta x_1 \Delta x_2 \Delta x_3)$ has been divided out of Equation 2.28. Equation 2.28 can be generalized to [4]

$$\sum_{j=1}^{3} \frac{\partial \sigma_{ij}}{\partial x_j} = \rho \frac{\partial^2 u_i}{\partial t^2} \tag{2.29}$$

Equations 2.29 are basic equations of elasticity, equivalent to Newton's second law, and will be used for several problems later, including the development of the elastic wave equation. The index i ranges from 1 to 3 to correspond to the three Cartesian components of the motion.

2.2.4 The Stress Tensor

The stress will be shown later to be a second-rank tensor. Equation 2.25 indicates that the tensor is symmetric for homogeneous stress. It will now be shown that Equation 2.25 holds rather generally, not just under the conditions of homogeneous stress.

Figure 2.12 illustrates the forces acting on a small volume element that produce torque about the x_1 axis [4]. The torque produced by the four forces illustrated is

$$\begin{aligned}&\left[\left(\sigma_{32} + \frac{\partial \sigma_{32}}{\partial x_2}\left(\frac{\Delta x_2}{2}\right)\right)(\Delta x_1 \Delta x_3) + \left(\sigma_{32} - \frac{\partial \sigma_{32}}{\partial x_2}\right)\left(\frac{\Delta x_2}{2}\right)(\Delta x_1 \Delta x_3)\right]\left(\frac{\Delta x_2}{2}\right)\\ &-\left[\left(\sigma_{23} + \frac{\partial \sigma_{23}}{\partial x_3}\left(\frac{\Delta x_3}{2}\right)\right)(\Delta x_1 \Delta x_2) + \left(\sigma_{23} - \frac{\partial \sigma_{23}}{\partial x_3}\left(\frac{\Delta x_3}{2}\right)\right)(\Delta x_1 \Delta x_2)\right]\left(\frac{\Delta x_3}{2}\right)\\ &= (\sigma_{32} - \sigma_{23})\Delta x_1 \Delta x_2 \Delta x_3.\end{aligned} \tag{2.30}$$

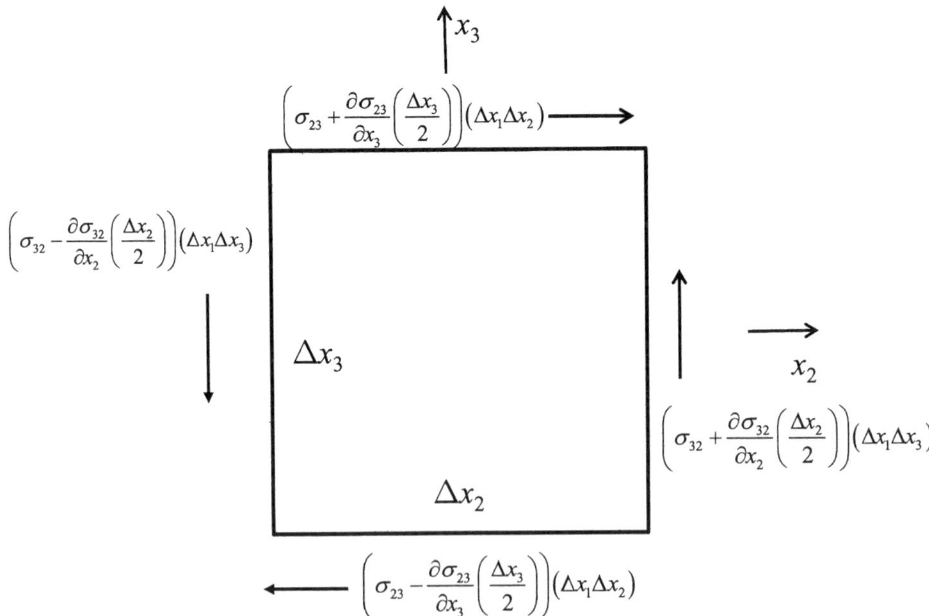

Figure 2.12 The stresses contributing to torque about the x_1 axis are shown. The x_1 axis is taken to be at the center of the square. The stresses are considered to vary spatially.

The result of Equation 2.30 is now used with Newton's second law of rotational motion, $\tau = I\alpha$, where τ is the torque, I is the moment of inertia, and α is the angular acceleration. For simplicity take $\Delta x_1 = \Delta x_2 = \Delta x_3 = \Delta x$. The moment of inertia of a cube rotating about the center as indicated in Figure 2.12 is $I = (1/6)m\Delta x^2 = 1/6\rho\Delta x^5$. With these simplifications, and using Equation 2.30, Newton's second law for rotational motion becomes,

$$(\sigma_{32} - \sigma_{23}) = \frac{\rho\alpha\Delta x^2}{6}. \tag{2.31}$$

Taking the limit as Δx goes to zero, shows that $\sigma_{32} = \sigma_{23}$. The only alternative is to have α approach infinity, which is unphysical. As there was nothing special for torques about the x_3 axis, it is concluded that

$$\sigma_{ij} = \sigma_{ji}. \tag{2.32}$$

There is also *experimental* evidence [10] that Equation 2.32 holds if there are no body torques. The present discussion has omitted body forces or body torques. Body torque effects could occur in materials with electric or magnetic moments. Even then, the effects are usually negligible in the linear approximation [11]. For the rest of this work, Equation 2.32 will be assumed valid.

The stress illustrated in Figure 2.9 forms a second-rank tensor, as will now be discussed in more detail [4, 11, 12]. The discussion is facilitated by reference to Figure 2.13. Figure 2.13 appears complicated, so it seems worthwhile to carefully define terms.

δS_i is an area of a face of the tetrahedron perpendicular to the x_i axis, *e.g.* δS_2 is the area of face *Oac*. There are three such faces.

δS_n is the area of the fourth face of the tetrahedron, *i.e. abc*. $\mathbf{n}$ is the normal to this face.

The force acting on *abc* is $\mathbf{f}\delta S_n$, *i.e.*, $\mathbf{f}$ has units of force per area.

The stresses σ_{ij} act on the three mutually perpendicular faces as indicated in Figure 2.13.

For the following analysis it is useful to define directions cosines for $\mathbf{n}$

$$\mathbf{n} = (l_1, l_2, l_3) \tag{2.33}$$

where l_i is the cosine of the angle between $\mathbf{n}$ and the x_i axis. It will also be useful to relate the areas defined in this section. Consideration of Figure 2.13 shows that δS_i is the projection of δS_n onto the plane perpendicular to the x_i axis. Further consideration shows that

$$\delta S_i = \delta S_n l_i. \tag{2.34}$$

Assuming static equilibrium, it is straightforward to relate $\mathbf{f}$ to $\mathbf{n}$. The relation will involve the σ_{ij} and illustrate the tensor nature of the stresses. Considering the x_1 component of the forces gives,

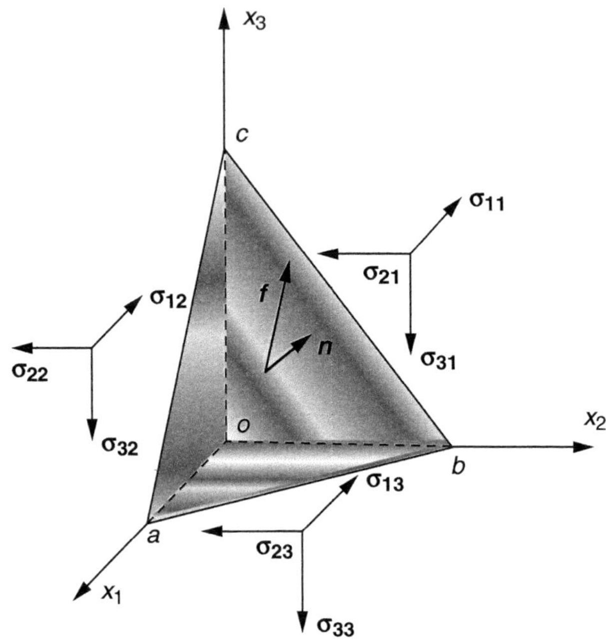

Figure 2.13 Forces acting on a tetrahedron are illustrated. It is assumed that these forces are exerted on the surface of the tetrahedron by the surrounding material, or by external forces. The stresses σ_{ij} act on the three mutually perpendicular faces, similar to those illustrated in Figure 2.9. A force, due to **f**, acts on the face defined by the triangle *abc*; **n** is a unit vector normal to *abc*. **f** is not necessarily parallel to **n**. It is convenient to take **f** as force per unit area.

$$f_1 \delta S_n = \sigma_{11} \delta S_1 + \sigma_{12} \delta S_2 + \sigma_{13} \delta S_3. \tag{2.35}$$

Equation 2.35 is simplified by using the results of Equation 2.34,

$$f_1 = \sigma_{11} l_1 + \sigma_{12} l_2 + \sigma_{13} l_3. \tag{2.36}$$

Similar results are obtained for the x_2 and x_3 directions. These results can be written generally as

$$\begin{pmatrix} f_1 \\ f_2 \\ f_3 \end{pmatrix} = \begin{pmatrix} \sigma_{11} & \sigma_{12} & \sigma_{13} \\ \sigma_{21} & \sigma_{22} & \sigma_{23} \\ \sigma_{31} & \sigma_{32} & \sigma_{33} \end{pmatrix} \begin{pmatrix} l_1 \\ l_2 \\ l_3 \end{pmatrix}. \tag{2.37}$$

Equation 2.37 shows that the σ_{ij} relate two vectors, **f** and **n**. ***This is just the property of a tensor of the second rank; the stress is a symmetric tensor of the second rank.*** (It is symmetric because $\sigma_{ij} = \sigma_{ji}$.)

Equation 2.37 relates forces on an arbitrarily oriented face to the Cartesian coordinate system stresses. Equation 2.37 is particularly useful if the face *abc* is part of the external surface. In that case, Equation 2.37 can be used to relate external forces to the internal stresses. ***This fact can be used to relate boundary conditions to the internal stresses.***

Simple Examples of Stresses

Similar to the discussion of Section 2.1.3, a relatively simple stress will be examined in some detail. Consider the stress tensor,

$$\sigma = \begin{pmatrix} 0 & \sigma & 0 \\ \sigma & 0 & 0 \\ 0 & 0 & 0 \end{pmatrix}. \tag{2.38}$$

The situation is illustrated in Figure 2.14a for the $x_1 - x_2$ plane, following the definition of stress in Figure 2.9. A pure shear stress is shown. It should be remembered that, for the general case of inhomogeneous stresses, the volumes illustrated are to be taken as infinitesimal. Equation 2.37 will now be used to examine the forces acting on a surface perpendicular to the x_1' axis, which is rotated 45° with respect to the x_1 axis. In this case, the unit vector $\mathbf{n}$ points along the x_1' axis. Equation 2.37 becomes

$$\begin{pmatrix} f_1 \\ f_2 \\ f_3 \end{pmatrix} = \begin{pmatrix} 0 & \sigma & 0 \\ \sigma & 0 & 0 \\ 0 & 0 & 0 \end{pmatrix} \begin{pmatrix} \frac{1}{\sqrt{2}} \\ \frac{1}{\sqrt{2}} \\ 0 \end{pmatrix}, \tag{2.39}$$

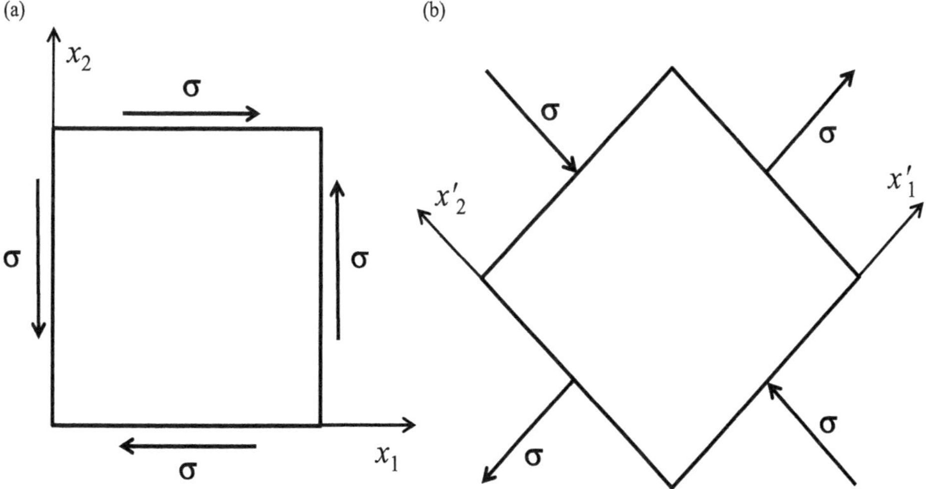

Figure 2.14 (a) Illustration of the stresses of Equation 2.38, which shows a pure shear stress in the unprimed coordinate system. Only the $x_1 - x_2$ plane is shown. (b) The same stress in the primed system, rotated 45° with respect to the original system. In this system there is a pure extensional stress and a pure compressional stress.

giving $f_1 = f_2 = \frac{\sigma}{\sqrt{2}}$. Recalling that f_i points along the original x_i axis, the result is that the force per area acting on the face perpendicular to the x_1' axis is $\sqrt{f_1^2 + f_2^2} = \sigma$ directed along the positive x_1' axis. For static equilibrium there must be a stress in the opposite direction on the opposite face of the cube.

Applying a similar argument gives $f_1 = \frac{\sigma}{\sqrt{2}} = -f_2$ for the stress on the face perpendicular to the x_2 axis, with a resultant stress of σ pointed along the negative x_2' axis. The situation is illustrated in Figure 2.14b. The pure shear stress of Figure 2.14a referred to the x_i axes is replaced by a compressional and extensional stress in Figure 2.14b referred to the x_i' axes.

Similiar to the discussion of strains in Section 2.1.3, the previous example of stress transformation can be treated by other means. The stress, being a second-rank tensor, transforms between coordinate systems exactly as does the strain, Equation 2.19,

$$\sigma_{ij}' = \sum_{kl} a_{ik} a_{jl} \sigma_{kl}. \tag{2.40}$$

Applying Equation 2.40 to the stress tensor of Equation 2.38 for a counterclockwise rotation of 45° gives

$$\sigma' = \begin{pmatrix} \sigma & 0 & 0 \\ 0 & -\sigma & 0 \\ 0 & 0 & 0 \end{pmatrix} \tag{2.41}$$

for the stress in the x_i' system. The stress tensor σ' of Equation 2.41 is precisely that illustrated in Figure 2.14b. Finally, another approach is to simply diagonalize the matrix of Equation 2.38 to find the principal axes and eigenvalues. This method gives the same result as Equation 2.41 for the eigenvalues, and shows the principal axes to be the x' axes of Figure 2.14b. Both the strain and stress results for these simple examples show how a pure shear in one coordinate system is represented as simple compressional or extensional effects in the principal axes system.

2.3 Elastic Constants

The study of the elasticity of materials is concerned with the deformations in response to applied forces. In previous sections deformations have been described in terms of strains, and applied forces in terms of stresses. Thus, a relation between stress and strain is needed (such relations are generally known as constitutive relations). It is well-known that, for sufficiently small deformations, the applied force and resulting deformation are linearly related; this effect is known as Hooke's Law. In the present language Hooke's Law might be written as $\sigma = c\epsilon$, where c

is a constant characteristic of the material. A generalized Hooke's Law describing this situation is

$$\sigma_{ij} = \sum_{kl} c_{ijkl}\epsilon_{kl}. \tag{2.42}$$

The indices *ijkl* run from 1 to 3 as usual. The alert reader will notice that there appear to be $3^4 = 81$ of the c_{ijkl}, a frightening thought. Fortunately, even in the most general case, there are many symmetry relations among the c_{ijkl} reducing the number of independent c_{ijkl} considerably. Further, in many cases of interest, many of the c_{ijkl} are zero. These simplifications will be discussed later.

The c_{ijkl} are known by various names: elastic constants, elastic moduli, or elastic stiffness constants. Previous sections have shown that both σ and ϵ are second rank tensors, thus, the c_{ijkl} form a fourth rank tensor [4].

To be explicit about the meaning of Equation 2.42, it will be written out in full for the case of a general σ_{ij} without taking into account the fact that many of the c_{ijkl} may be zero for any particular case. The resulting equation is

$$\begin{aligned}\sigma_{ij} = c_{ij11}\epsilon_{11} + c_{ij12}\epsilon_{12} + c_{ij13}\epsilon_{13}\\ + c_{ij21}\epsilon_{21} + c_{ij22}\epsilon_{22} + c_{ij23}\epsilon_{23}\\ + c_{ij31}\epsilon_{31} + c_{ij32}\epsilon_{32} + c_{ij33}\epsilon_{33}.\end{aligned} \tag{2.43}$$

In many cases it is convenient to express the relation between stress and strain as

$$\epsilon_{ij} = \sum_{kl} s_{ijkl}\sigma_{kl}, \tag{2.44}$$

where the s_{ijkl} are called the elastic compliances. As was the case for the c_{ijkl}, the s_{ijkl} form a fourth-rank tensor.

2.3.1 Transformation of Elastic Constants between Coordinate Systems

In many cases it proves useful to transform the elastic constants from the x_i system to the x'_i system where the two systems are related by direction cosines as in Equation 2.19. Because the elastic constants form a fourth-rank tensor they transform as [4],

$$c'_{ijkl} = \sum_{mnop} a_{im}a_{jn}a_{ko}a_{lp}c_{mnop}. \tag{2.45}$$

Just as in Equation 2.19, a_{im} is a direction cosine where the first subscript refers to the x'_i axis while the second refers to the x_m axis. Equation 2.45 is rather tedious to use by hand, but may be readily programmed – including symbolically – for use with various commercially available software programs.

Table 2.1 *Voigt notation.*

Tensor Notation	11	22	33	23,32	13,31	12,21
Voigt Notation	1	2	3	4	5	6

Table 2.2 *Relation between tensor and matrix strains.*

Tensor Notation	ϵ_{11}	ϵ_{22}	ϵ_{33}	$\epsilon_{32},\epsilon_{23}$	$\epsilon_{13},\epsilon_{31}$	$\epsilon_{12},\epsilon_{21}$
Matrix Notation	ϵ_1	ϵ_2	ϵ_3	$\epsilon_4/2$	$\epsilon_5/2$	$\epsilon_6/2$

2.3.2 Voigt Notation

From the definitions of strain, Equation 2.9, $\epsilon_{ij} = \epsilon_{ji}$. Also, if no body torques act, $\sigma_{kl} = \sigma_{lk}$ (Equation 2.32). This symmetry under the interchange of indices make possible a considerable simplification with the change in notation shown in Table 2.1, often known as the Voigt notation. Thus, ϵ_{11} becomes ϵ_1 and c_{1131} becomes c_{15}, *etc.* However, the situation is a bit more complicated, and certain conventions have been established, as will now be explored in more detail. Consider again Equation 2.43. Because $\epsilon_{12} = \epsilon_{21}$ identically, the stress produced by ϵ_{12} must be identical to that produced by ϵ_{21}, so it must be that $c_{ij12} = c_{ij21}$, or in general

$$c_{ijkl} = c_{ijlk}. \tag{2.46}$$

Further insight is obtained by writing an equation for σ_{ji}, similar to Equation 2.43, and remembering that $\sigma_{ij} = \sigma_{ji}$, Equation 2.32. The result is that it must be that $c_{ij12} = c_{ji12}$, or in general

$$c_{ijkl} = c_{jikl}. \tag{2.47}$$

With these results, Equation 2.43 for σ_{11} may be rewritten

$$\sigma_1 = c_{11}\epsilon_1 + c_{16}\epsilon_6 + c_{15}\epsilon_5 + c_{12}\epsilon_2 + c_{14}\epsilon_4 + c_{13}\epsilon_3 \tag{2.48}$$

where the ϵ_m are defined in Table 2.2. Note the factor $1/2$ in the table, which follows automatically from the symmetry of the strains, $\epsilon_{ij} = \epsilon_{ji}$, and the definitions of the two index elastic moduli, $c_{ijkl} = c_{mn}$, with m and n being determined according to Table 2.1[1]. (There are other possibilities for the relations between tensor and matrix strains, but then the relation between the tensor c_{ijkl} and the matrix c_{mn} would be different. Table 2.2 is the conventional choice, [4, 13] but the reader should be aware that some earlier writers [12] used a different choice.)

[1] $\sigma_{ij} = \sigma_m$ by Table 2.1 with no factors of 2.

Table 2.3 *Matrix-tensor relations for compliances and elastic constants.*

Relation	Condition
$s_{ijkl} = s_{mn}$	Both m and n are 1, 2, or 3
$2s_{ijkl} = s_{mn}$	Either m or n, but not both, are 4, 5, or 6
$4s_{ijkl} = s_{mn}$	Both m and n are 4, 5, or 6
$c_{ijkl} = c_{mn}$	For any m and n = 1 ... 6

A change from tensor to matrix notation is also possible for the s_{ijkl}; however, the result is a bit more complicated than was the case for the c_{ijkl} because of the factor 1/2 in Table 2.2. The results can be obtained by writing out an equation similar to Equation 2.43 for ϵ_{ii} for $i = 1, 2$, or 3.

$$\begin{aligned}\epsilon_{ii} &= s_{ii11}\sigma_{11} + s_{ii12}\sigma_{12} + s_{ii13}\sigma_{13} \\ &\quad + s_{ii21}\sigma_{21} + s_{ii22}\sigma_{22} + s_{ii23}\sigma_{23} \\ &\quad + s_{ii31}\sigma_{31} + s_{ii32}\sigma_{32} + s_{ii33}\sigma_{33}.\end{aligned} \tag{2.49}$$

Considering that, *e.g.* $\sigma_{12} = \sigma_{21}$, then each of these stresses must contribute equally to ϵ_{ii}; thus $s_{ii12} = s_{ii21}$, *etc.* Taking *e.g.* $i = 1$, Equation 2.49 may be rewritten as

$$\epsilon_1 = s_{11}\sigma_1 + s_{16}\sigma_6 + s_{15}\sigma_5 + s_{12}\sigma_2 + s_{14}\sigma_4 + s_{13}\sigma_3 \tag{2.50}$$

where $s_{11} = s_{1111}$ and $s_{16} = 2s_{1112}$, *etc.*

An equation for ϵ_{ij} for $i \neq j$ is also needed. To be specific, taking $(ij) = (12)$, proceeding as for Equation 2.49, and noting the factor 1/2 in Table 2.2 for ϵ_{12} gives

$$\epsilon_6 = s_{61}\sigma_1 + s_{66}\sigma_6 + s_{65}\sigma_5 + s_{62}\sigma_2 + s_{44}\sigma_4 + s_{63}\sigma_3 \tag{2.51}$$

where $s_{66} = 4s_{1212}$ and $s_{61} = 2s_{1211}$, *etc.*

To summarize, the complete relations between the s_{ijkl} and the s_{mn} are given in Table 2.3. The simple relation between tensor and matrix elastic constants is included in the table.

Using the condensed notation Hooke's Law, Equation 2.42 becomes

$$\sigma_m = \sum_{n=1}^{6} c_{mn}\epsilon_n. \tag{2.52}$$

The number of independent c_{mn} is 36, and the same holds true for the number of independent s_{mn}. This is a significant reduction from the number, 81, deduced from Equation 2.42. More reductions follow after development of some necessary material.

2.3.3 Relation of Elastic Constants to Thermodynamics

Work Required for Infinitesimal Deformations

It will be useful to have expressions for the energy associated with elastic deformations. The work required to produce an infinitesimal change in strain will be calculated. It is assumed that $\epsilon_{ij} \to \epsilon_{ij} + d\epsilon_{ij}$. Then $\Delta u_i \to \Delta u_i + du_i$. Making these substitutions in Equations 2.18 leads to,

$$\begin{pmatrix} du_1 \\ du_2 \\ du_3 \end{pmatrix} = \begin{pmatrix} d\epsilon_{11} & d\epsilon_{12} & d\epsilon_{13} \\ d\epsilon_{21} & d\epsilon_{22} & d\epsilon_{23} \\ d\epsilon_{31} & d\epsilon_{32} & d\epsilon_{33} \end{pmatrix} \begin{pmatrix} \Delta x_1 \\ \Delta x_2 \\ \Delta x_3 \end{pmatrix}. \tag{2.53}$$

where the *changes* in strain $d\epsilon_{ij}$, produce the du_i at Δx_j.

As a specific example the work associated with producing an increase in strain ϵ_{11} will be considered. All other strains are zero. The work will be performed by the applied stress σ_{11}. In general, other stresses will be required to maintain all strains except ϵ_{11} zero, but these other stresses do no work, because the associated strains are zero. The situation is illustrated in Figure 2.15. The change in strain $d\epsilon_{11}$ is not shown in the figure. The infinitesimal work performed by σ_{11} is

$$\begin{aligned} \delta W_{11} &= du_1 F_1 \\ &= [d\epsilon_{11}\Delta\mathcal{L}_1/2][\sigma_{11}(\Delta\mathcal{L}_2\Delta\mathcal{L}_3)] + [d\epsilon_{11}(-\Delta\mathcal{L}_1/2)][-\sigma_{11}(\Delta\mathcal{L}_2\Delta\mathcal{L}_3)] \\ &= [d\epsilon_{11}\Delta\mathcal{L}_1][\sigma_{11}(\Delta\mathcal{L}_2\Delta\mathcal{L}_3)] \\ &= d\epsilon_{11}\sigma_{11}V. \end{aligned} \tag{2.54}$$

The second line in Equation 2.54 has two terms: one for the front face of the cube; and, one for the back face. The sign of the displacement and the sign of the force

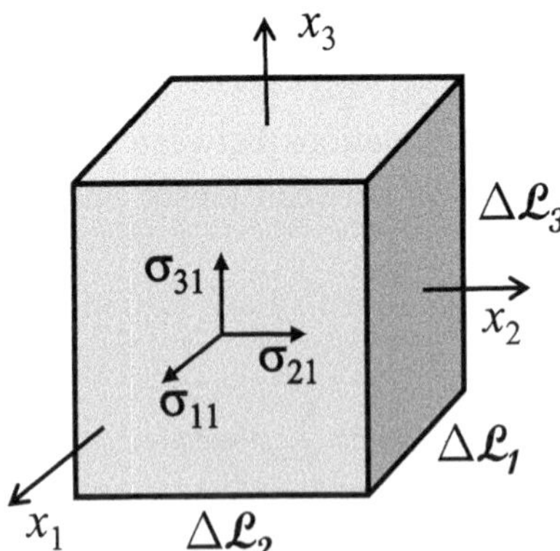

Figure 2.15 An illustration of an infinitesimal cube of material. Stresses are applied so as to cause a change in the strain ϵ_{11}. Other strains do not change. Work is done by the stress σ_{11}.

are reversed on the back face, but in combination they give a positive contribution to the work. Δx_1 of Equation 2.53 becomes $\pm\Delta\mathcal{L}_1/2$ in Equation 2.54. V is the volume of the cube.

Other stresses are handled in the same manner. For example, the work performed by the stress σ_{21} when all strains except ϵ_{12} are held constant is

$$\begin{aligned}\delta W_{21} &= du_2 F_2 \\ &= [d\epsilon_{21}\Delta\mathcal{L}_1/2][\sigma_{21}(\Delta\mathcal{L}_2\Delta\mathcal{L}_3)] + [d\epsilon_{21}(-\Delta\mathcal{L}_1/2)][-\sigma_{11}(\Delta\mathcal{L}_2\Delta\mathcal{L}_3)] \\ &= [d\epsilon_{21}\Delta\mathcal{L}_1][\sigma_{11}(\Delta\mathcal{L}_2\Delta\mathcal{L}_3)] \\ &= d\epsilon_{21}\sigma_{21}V. \end{aligned} \tag{2.55}$$

One could imagine proceeding in a similiar manner to compute the work required to produce an increase in each of the nine strains in the strain matrix of Equation 2.53. In the following it is assumed that δW is the work *per unit volume* and the V will be suppressed. Adding together all the contributions gives the infinitesimal work as

$$\delta W = \sum_{i,j} \sigma_{ij} d\epsilon_{ij}, \tag{2.56}$$

Adiabatic Elastic Constants

The first law of thermodynamics is

$$dU = \delta Q + \delta W \tag{2.57}$$

where dU is the change in the internal energy of the system, δW is the work done on the system, and δQ is the heat flow into the system. As usual, δ indicates an infinitesimal quantity, but d indicates a differential. The difference arises because U is a thermodynamic state function, but W and Q are not. For reversible processes, the second law of thermodynamics gives

$$\delta Q = TdS, \tag{2.58}$$

leading to

$$dU = TdS + \delta W \tag{2.59}$$

where S is the entropy and T is the temperature.

For adiabatic processes, there is no heat exchange between the system and the surroundings – or between different parts of the system – so $\delta Q = 0$ and $dS = 0$ with the result that $dU = \delta W$. Then, using Equation 2.56 gives

$$dU = \sum_{i,j} \sigma_{ij} d\epsilon_{ij}. \tag{2.60}$$

which leads to the important result

$$\sigma_{ij} = \left(\frac{\partial U}{\partial \epsilon_{ij}}\right)_S. \tag{2.61}$$

Combining Equations 2.42 and 2.61 gives

$$c^S_{ijkl} = \left(\frac{\partial \sigma_{ij}}{\partial \epsilon_{kl}}\right)_S = \left(\frac{\partial^2 U}{\partial \epsilon_{ij}\partial \epsilon_{kl}}\right)_S = \left(\frac{\partial^2 U}{\partial \epsilon_{kl}\partial \epsilon_{ij}}\right)_S = c^S_{klij}. \tag{2.62}$$

The order of the differentiation of U does not matter, which permits the interchange of indices: $c^S_{ijkl} = c^S_{klij}$. The superscript S indicates constant entropy, *i.e.* adiabatic, elastic constants. Equation 2.62 shows a fundamental relation: *the adiabatic elastic constants are the second derivatives of the internal energy with respect to strain.* The adiabatic elastic constants are the ones usually measured in ultrasonic experiments.

Isothermal Elastic Constants

The Helmholtz free energy is defined as $F = U - TS$. Using this relation to replace U in Equation 2.59 it is easily shown that $dF = \delta W$ *for isothermal processes.* Results similar to Equations 2.60 to 2.62 immediately follow with dF replacing dU and the subscript T replacing the subscript S so that

$$c^T_{ijkl} = \left(\frac{\partial \sigma_{ij}}{\partial \epsilon_{kl}}\right)_T = \left(\frac{\partial^2 F}{\partial \epsilon_{ij}\partial \epsilon_{kl}}\right)_T = \left(\frac{\partial^2 F}{\partial \epsilon_{kl}\partial \epsilon_{ij}}\right)_T = c^T_{klij}. \tag{2.63}$$

The isothermal elastic constants are the second derivatives of the Helmholtz free energy with respect to strain. The isothermal elastic constants are the ones commonly derived theoretically. The superscript T indicates isothermal elastic constants. The superscripts S and T will be used when it is important to distinguish between adiabatic and isothermal quantities.

Reduction of the Number of Independent Elastic Constants to 21

The relations expressed by Equations 2.46 and 2.47 allowed the introduction of the Voigt notation, which reduced the number of independent elastic constants from 81 to 36. Equation 2.62, or Equation 2.63, permit a further reduction. In the reduced notation these equations have the form $c_{mn} = c_{nm}$. In this notation Equation 2.52 becomes

$$\begin{pmatrix} \sigma_1 \\ \sigma_2 \\ \sigma_3 \\ \sigma_4 \\ \sigma_5 \\ \sigma_6 \end{pmatrix} = \begin{pmatrix} c_{11} & c_{12} & c_{13} & c_{14} & c_{15} & c_{16} \\ c_{12} & c_{22} & c_{23} & c_{24} & c_{25} & c_{26} \\ c_{13} & c_{23} & c_{33} & c_{34} & c_{35} & c_{36} \\ c_{14} & c_{24} & c_{34} & c_{44} & c_{45} & c_{46} \\ c_{15} & c_{25} & c_{35} & c_{45} & c_{55} & c_{56} \\ c_{16} & c_{26} & c_{36} & c_{46} & c_{56} & c_{66} \end{pmatrix} \begin{pmatrix} \epsilon_1 \\ \epsilon_2 \\ \epsilon_3 \\ \epsilon_4 \\ \epsilon_5 \\ \epsilon_6 \end{pmatrix}. \tag{2.64}$$

There are 21 independent values of c_{mn} in Equation 2.64. This is the most general form of the elastic constant matrix which holds for crystals of triclinic symmetry. As will be discussed later, crystals of higher symmetry have fewer independent elastic constants. It is important to remember that the elastic constant matrix appearing in Equation 2.64 is not a tensor, and does not transform between coordinate systems as a tensor.

Elastic Energy

It is useful to have an expression for the elastic energy density, and for that purpose Equation 2.60 is a convenient starting point. Using the generalized Hooke's Law for σ_{ij}, Equation 2.42, and switching to the Voigt notation results in

$$\begin{aligned} dU = \sum_{mn} c_{mn}\epsilon_m d\epsilon_n = \\ c_{11}\epsilon_1 d\epsilon_1 + c_{12}\epsilon_1 d\epsilon_2 + c_{13}\epsilon_1 d\epsilon_3 + \cdots c_{16}\epsilon_1 d\epsilon_6 + \\ c_{21}\epsilon_2 d\epsilon_1 + c_{22}\epsilon_2 d\epsilon_2 + c_{23}\epsilon_2 d\epsilon_3 + \cdots c_{26}\epsilon_2 d\epsilon_6 + \\ \cdots\cdots\cdots\cdots\cdots\cdots + \\ \cdots\cdots\cdots\cdots\cdots\cdots + \\ \cdots\cdots\cdots\cdots\cdots\cdots + \\ c_{61}\epsilon_6 d\epsilon_1 + c_{62}\epsilon_6 d\epsilon_2 + \cdots c_{66}\epsilon_6 d\epsilon_6 \end{aligned} \tag{2.65}$$

where only a few of the 36 terms are written out explicitly. Equation 2.65 may be utilized to get the energy U. Note that Equation 2.65 may be considered the total differential of U for infinitesimal changes in strain, *i.e.*,

$$dU = \left(\frac{\partial U}{\partial \epsilon_1}\right)_{\epsilon_{i\neq 1}} d\epsilon_1 + \left(\frac{\partial U}{\partial \epsilon_2}\right)_{\epsilon_{i\neq 2}} d\epsilon_2 + \cdot\cdot + \cdot\cdot + \cdot\cdot + \left(\frac{\partial U}{\partial \epsilon_6}\right)_{\epsilon_{i\neq 6}} d\epsilon_6. \tag{2.66}$$

Comparing the two equations gives,

$$\left(\frac{\partial U}{\partial \epsilon_1}\right)_{\epsilon_{i\neq 1}} = c_{11}\epsilon_1 + c_{21}\epsilon_2 + c_{31}\epsilon_3 + c_{41}\epsilon_4 + c_{51}\epsilon_5 + c_{61}\epsilon_6, \tag{2.67}$$

with similar expressions for the other partial derivatives. One contribution to U is found by integrating Equation 2.67 over ϵ_1 keeping the other ϵ_i constant. Doing the same thing for the other five ϵ_i and adding the results gives

$$U = \frac{1}{2}c_{11}\epsilon_1^2 + \frac{1}{2}c_{22}\epsilon_2^2 \cdots \frac{1}{2}c_{66}\epsilon_6^2 + c_{12}\epsilon_1\epsilon_2 + c_{13}\epsilon_1\epsilon_3 + \cdots c_{56}\epsilon_5\epsilon_6, \tag{2.68}$$

where the constant of integration is taken as 0. Only a few of the 36 terms on the right-hand side of Equation 2.68 are shown. Equation 2.68 can be written much more elegantly as [4]

$$U = \frac{1}{2} \sum_{mn} c_{mn} \epsilon_m \epsilon_n, \tag{2.69}$$

the well-known equation for the elastic energy density. In applications of Equation 2.69 it is useful to remember that $c_{mn} = c_{nm}$. The strains, and in principle the elastic constants, could be functions of position. It would then be necessary to integrate U over the volume of the specimen to obtain the total elastic strain energy. The *adiabatic* elastic constants appear in Equation 2.69. A similar expression holds for the elastic free energy F, in which case the *isothermal* elastic constants would appear.

2.3.4 Elastic Constants and Symmetry

Quite general considerations have shown that, if body torques are zero or insignificant, the number of independent elastic constants is reduced to 21 (Equation 2.64). Further reductions are possible, depending on symmetry [6]. Point symmetry operations will be applied to the matrix of Equation 2.64 to achieve these reductions. For the present discussion the only point symmetry operations to be considered are: inversion through a point, denoted by $\bar{1}$; reflection in a plane, denoted by (m); and an n-fold rotation (rotation by $2\pi/n$ about a point), denoted by n.

The inversion operation has the following result,

$$x_1 \to -x_1, x_2 \to -x_2, x_3 \to -x_3. \tag{2.70}$$

It follows that $u_1 \to -u_1, u_2 \to -u_2, u_3 \to -u_3$.

As an example of m consider a mirror reflection in a plane perpendicular to the x_1 axis

$$x_1 \to -x_1, x_2 \to x_2, x_3 \to x_3. \tag{2.71}$$

Finally, as an example of a rotation, consider a four-fold rotation about the x_3 axis, *i.e.* $n = 4$,

$$x_1 \to x_2, x_2 \to -x_1, x_3 \to x_3. \tag{2.72}$$

Next, the effects of these operations on the elastic constants, c_{ijkl}, will be considered. First, recalling the definition of strain from Equation 2.9, it is seen that if both x_i and x_j change sign, then ϵ_{ij} *does not* change sign, but if only one of these coordinates undergoes a sign change, then ϵ_{ij} *reverses* sign. For example, considering the mirror reflection of Equation 2.71 it is seen that this operation produces the following: $\epsilon_{12} \to -\epsilon_{12}$; but, $\epsilon_{11} \to +\epsilon_{11}$. Equation 2.62 shows that a change in the sign of a strain produces a change in the sign of the associated c_{ijkl}. It will be most useful to consider the case of reflections in a plane. Suppose that a particular axis, *e.g.*, p, is perpendicular to a reflection plane. Then, if p appears an odd number of times in c_{ijkl}, the mirror reflection reverses the sign of c_{ijkl}, but if it appears an even number of times there is no sign change [6]. Considering the example of Equation 2.71 (the mirror plane is perpendicular to the x_1 axis) it is seen that $c_{1233} \rightarrow -c_{1233}$, but $c_{1133} \rightarrow +c_{1133}$.

It is clear from the above discussion that an inversion, Equation 2.70, causes a change in any c_{ijkl} of $(-1)^4$ and thus has no effect.

A two-fold rotation about any axis plus an inversion gives the same result as a reflection in a plane perpendicular to the chosen axis. For example, combining a two-fold rotation about the x_1 axis ($x_1 \rightarrow x_1, x_2 \rightarrow -x_2, x_3 \rightarrow -x_3$) with an inversion leads to the net effect of $x_1 \rightarrow -x_1, x_2 \rightarrow x_2, x_3 \rightarrow x_3$, which is the same effect produced by a reflection in a plane perpendicular to x_1. However, the inversion has no effect on the c_{ijkl}, so for effects on the c_{ijkl} of a two-fold rotation for a system with an inversion center, only the reflection need be considered, which is usually a simpler approach.

The Seven Crystal Systems

The effect of symmetry on the number of independent elastic constants will now be considered [4, 6, 14]. Starting with the general elastic constant matrix, Equation 2.64, the point group symmetries of the different crystal classes will be applied to determine relations among the different c_{ijkl} required by symmetry. The discussion will illustrate the use of symmetry arguments, but will not be exhaustive. Results are given in Tables 2.4 and 2.5.

TRICLINIC The lowest symmetry crystal is the triclinic system. Apart from the identity operation, this system has only inversion symmetry, but inversion imposes no restrictions on the elastic constants. Thus, the elastic constant matrix for this case has 21 independent elements.

MONOCLINIC The monoclinic system has a two-fold rotation axis, or a mirror plane, or a two-fold rotation axis with a mirror plane perpendicular to it, which means that the physical properties of the system are invariant under these operations. As discussed earlier, a two-fold rotation axes plus an inversion is equal to a reflection perpendicular to the axis, but the inversion has no effect on the elastic constants. This means that the symmetry effects on the monoclinic system can be analyzed in terms of a reflection perpendicular to the rotation axis. A reflection involves a sign change. If $c_{ijkl} = -c_{ijkl}$, then the elastic constant must $= 0$ for the system to be invariant under the operation. There appear to be two common ways to describe the monoclinic system: the unique axis parallel to x_2; or, the unique axis parallel to x_3. The following discussion will be for x_2 along the unique axis. It follows that any c_{ijkl} for which the subscript 2 appears an odd number of times must be zero. A little thought shows that such constants as $c_{14} = c_{1123}, c_{16} = c_{1112}, c_{24} = c_{2223}, c_{26} = c_{2212}, c_{45} = c_{2313}$ *etc.* $= 0$, but $c_{15} = c_{1113}, c_{25} = c_{2213}$ *etc.* are not required by symmetry to be zero. Continuing this approach for all the c_{mn} results in

Table 2.4 *Elastic constant matrices for the different crystal systems I.*

Triclinic

$$\begin{pmatrix} c_{11} & c_{12} & c_{13} & c_{14} & c_{15} & c_{16} \\ c_{12} & c_{22} & c_{23} & c_{24} & c_{25} & c_{26} \\ c_{13} & c_{23} & c_{33} & c_{34} & c_{35} & c_{36} \\ c_{14} & c_{24} & c_{34} & c_{44} & c_{45} & c_{46} \\ c_{15} & c_{25} & c_{35} & c_{45} & c_{55} & c_{56} \\ c_{16} & c_{26} & c_{36} & c_{46} & c_{56} & c_{66} \end{pmatrix}$$

Twenty-one independent elastic constants

Orthorhombic

$$\begin{pmatrix} c_{11} & c_{12} & c_{13} & 0 & 0 & 0 \\ c_{12} & c_{22} & c_{23} & 0 & 0 & 0 \\ c_{13} & c_{23} & c_{33} & 0 & 0 & 0 \\ 0 & 0 & 0 & c_{44} & 0 & 0 \\ 0 & 0 & 0 & 0 & c_{55} & 0 \\ 0 & 0 & 0 & 0 & 0 & c_{66} \end{pmatrix}$$

Nine independent elastic constants

Monoclinic

$$\begin{pmatrix} c_{11} & c_{12} & c_{13} & 0 & c_{15} & 0 \\ c_{12} & c_{22} & c_{23} & 0 & c_{25} & 0 \\ c_{13} & c_{23} & c_{33} & 0 & c_{35} & 0 \\ 0 & 0 & 0 & c_{44} & 0 & c_{46} \\ c_{15} & c_{25} & c_{35} & 0 & c_{55} & 0 \\ 0 & 0 & 0 & c_{46} & 0 & c_{66} \end{pmatrix}$$

Thirteen independent elastic constants
x_2 is the unique axis

Tetragonal
Point symmetries $4, \overline{4}, \frac{4}{m}$

$$\begin{pmatrix} c_{11} & c_{12} & c_{13} & 0 & 0 & c_{16} \\ c_{12} & c_{11} & c_{13} & 0 & 0 & -c_{16} \\ c_{13} & c_{13} & c_{33} & 0 & 0 & 0 \\ 0 & 0 & 0 & c_{44} & 0 & 0 \\ 0 & 0 & 0 & 0 & c_{44} & 0 \\ c_{16} & -c_{16} & 0 & 0 & 0 & c_{66} \end{pmatrix}$$

Seven independent elastic constants

Monoclinic

$$\begin{pmatrix} c_{11} & c_{12} & c_{13} & 0 & 0 & c_{16} \\ c_{12} & c_{22} & c_{23} & 0 & 0 & c_{26} \\ c_{13} & c_{23} & c_{33} & 0 & 0 & c_{36} \\ 0 & 0 & 0 & c_{44} & c_{45} & 0 \\ 0 & 0 & 0 & c_{45} & c_{55} & 0 \\ c_{16} & c_{26} & c_{36} & 0 & 0 & c_{66} \end{pmatrix}$$

Thirteen independent elastic constants
x_3 is the unique axis

Tetragonal
Point symmetries $422, 4mm, \overline{4}2m, \frac{4}{m}\frac{2}{m}\frac{2}{m}$

$$\begin{pmatrix} c_{11} & c_{12} & c_{13} & 0 & 0 & 0 \\ c_{12} & c_{11} & c_{13} & 0 & 0 & 0 \\ c_{13} & c_{13} & c_{33} & 0 & 0 & 0 \\ 0 & 0 & 0 & c_{44} & 0 & 0 \\ 0 & 0 & 0 & 0 & c_{44} & 0 \\ 0 & 0 & 0 & 0 & 0 & c_{66} \end{pmatrix}$$

Six independent elastic constants

the elastic constant matrix of Table 2.4. The axis x_2 must lie along the crystalline unique axis for the matrix to have the given form; the orientation of the other two axes is not restricted.

In many cases the unique axis is taken to be parallel to c_3. In this case any c_{ijkl} for which the subscript 3 appears an odd number of times must be zero. The result is given in the table. The axis x_3 must lie along the crystalline unique axis for the matrix to have the given form; the orientation of the other two axes is not restricted.

ORTHORHOMBIC The three orthorhombic point groups have three mutually perpendicular two-fold rotation axes, three mutually perpendicular mirror planes, or a combination of the two. The x_1, x_2, and x_3 axes will be taken to lie along the two-fold rotation axes, or along the normals to the mirror planes. This means that any c_{ijkl} for which any subscript, 1, 2, or 3, appears an odd number of times must be zero. Applying this condition systematically to the c_{ijkl} leads immediately to the orthorhombic elastic constant matrix of Table 2.4. The x_i axes lie along the three, two-fold rotation axes.

TETRAGONAL Three of the tetragonal point groups are 4, $\overline{4}$, and $\frac{4}{m}$. The four-fold rotation includes the two-fold rotation, so initially the case is treated as for the monoclinic case with the unique axis parallel to x_3 [6]. The result of this initial step is the monoclinic elastic constant matrix for the unique axis parallel to x_3, Table 2.4. The result of a four-fold rotation is given by Equation 2.72. Applying this equation gives $c_{16} = c_{1112} \rightarrow -c_{2221} = -c_{26}$. Similar considerations show that $c_{36} = c_{3312} \rightarrow -c_{3321} = -c_{36} = 0$. The four-fold rotation about the x_3 axis also results in $c_{45} = 0$. The result for the tetragonal elastic constant matrix is given in Table 2.4, where the x_3 axis points along the crystalline unique axis.

All the other tetragonal point groups have mirror planes, or equivalently for the present argument, two-fold rotation axes perpendicular to the x_3 axis. This means that if the subscript 1 or 2 appears an odd number of times, the corresponding elastic constant is zero. In particular, $c_{16} = c_{1112}$ is zero. Applying this result gives the second tetragonal elastic constant of Table 2.4, where the x_3 axis points along the crystalline unique axis and the other two axes point along two-fold rotation axes or along the perpendiculars to mirror planes.

TRIGONAL The methods of the previous cases cannot be used for the trigonal case, because the simple relations of Equation 2.72 do not apply for three-fold rotations. An analytical approach in needed [6]. All the trigonal classes have a three-fold rotation axis; the physical properties are invariant under such rotations. Equation 2.45 will be used for rotations about the x_3 axis, which is taken to lie along the three-fold rotation axis. Requiring $c_{ijkl}(0) = c_{ijkl}(2\pi/3) = c_{ijkl}(4\pi/3)$ will place symmetry-related restrictions on the trigonal elastic constant tensor. For example it is found that

$$c_{1212}\left(\frac{2\pi}{3}\right) - c_{1212}\left(\frac{4\pi}{3}\right) = \frac{\sqrt{3}}{2}\left(c_{2212} - c_{1112}\right). \tag{2.73}$$

Because the left-hand side of Equation 2.73 must equal zero, the result is $c_{2212} = c_{1112}$, or $c_{26} = c_{16}$. Using a second relation,

$$c_{1111}\left(\frac{2\pi}{3}\right) - c_{1111}\left(\frac{4\pi}{3}\right) = -\frac{\sqrt{3}}{2}\left(3c_{2212} + c_{1112}\right). \tag{2.74}$$

The left-hand side of Equation 2.74 must equal zero. Combining the results of Equations 2.73 and 2.74 gives

$$c_{16} = c_{26} = 0, \tag{2.75}$$

where the notation switches between the two-index and four-index cases.

Table 2.5 *Elastic constant matrices for the different crystal systems II.*

Trigonal

Point symmetries 3, $\overline{3}$

$$\begin{pmatrix} c_{11} & c_{12} & c_{13} & c_{14} & -c_{25} & 0 \\ c_{12} & c_{11} & c_{13} & -c_{14} & c_{25} & 0 \\ c_{13} & c_{13} & c_{33} & 0 & 0 & 0 \\ c_{14} & -c_{14} & 0 & c_{44} & 0 & c_{25} \\ -c_{25} & c_{25} & 0 & 0 & c_{44} & c_{14} \\ 0 & 0 & 0 & c_{25} & c_{14} & \frac{1}{2}(c_{11}-c_{12}) \end{pmatrix}$$

Seven independent elastic constants

Trigonal

Point symmetries $3m$, 32, $\overline{3}\frac{2}{m}$

$$\begin{pmatrix} c_{11} & c_{12} & c_{13} & c_{14} & 0 & 0 \\ c_{12} & c_{11} & c_{13} & -c_{14} & 0 & 0 \\ c_{13} & c_{13} & c_{33} & 0 & 0 & 0 \\ c_{14} & -c_{14} & 0 & c_{44} & 0 & 0 \\ 0 & 0 & 0 & 0 & c_{44} & c_{14} \\ 0 & 0 & 0 & 0 & c_{14} & \frac{1}{2}(c_{11}-c_{12}) \end{pmatrix}$$

Six independent elastic constants

The x_1 axis is chosen as parallel to the two-fold rotation axis,or perpendicular to the mirror plane.

Hexagonal

$$\begin{pmatrix} c_{11} & c_{12} & c_{13} & 0 & 0 & 0 \\ c_{12} & c_{11} & c_{13} & 0 & 0 & 0 \\ c_{13} & c_{13} & c_{33} & 0 & 0 & 0 \\ 0 & 0 & 0 & c_{44} & 0 & 0 \\ 0 & 0 & 0 & 0 & c_{44} & 0 \\ 0 & 0 & 0 & 0 & 0 & \frac{1}{2}(c_{11}-c_{12}) \end{pmatrix}$$

Five independent elastic constants

Cubic

$$\begin{pmatrix} c_{11} & c_{12} & c_{12} & 0 & 0 & 0 \\ c_{12} & c_{11} & c_{12} & 0 & 0 & 0 \\ c_{12} & c_{12} & c_{11} & 0 & 0 & 0 \\ 0 & 0 & 0 & c_{44} & 0 & 0 \\ 0 & 0 & 0 & 0 & c_{44} & 0 \\ 0 & 0 & 0 & 0 & 0 & c_{44} \end{pmatrix}$$

Three independent elastic constants

Continuing with transformations similar to those of Equations 2.73 and Equation 2.74 the number of independent elastic constants is reduced from 21 (the triclinic case) to seven, as shown in Table 2.5 for point groups 3 and $\overline{3}$. The x_3 axis points along the three-fold rotation axis, but the orientation of the other axes is arbitrary. (Of course the actual values of the various elastic constants will depend on the choice of coordinate systems, it is only necessary for the x_3 axis to point along the three-fold rotation axis for the matrix to have the form shown.)

The other trigonal point groups have either a two-fold rotation axis perpendicular to the three-fold axis, or a mirror plane parallel to the three-fold axis. In this case the x_1 axis is chosen as parallel to the two-fold rotation axis, or perpendicular to the mirror plane. Following the discussion presented earlier it follows that $c_{25} = c_{2213} = 0$, resulting in the second elastic constant matrix for trigonal in Table 2.5. The second entry in Table 2.5 will have a different form if the x_2 axis is chosen as the special axis perpendicular to the x_3 axis.

For both the tetragonal and trigonal cases, it is possible to reduce the seven elastic constant cases to six elastic constants by choosing a coordinate system rotated by an appropriate angle β about the x_3 axis, but the correct value of β depends on the elastic constants [6]. It seems a matter of taste or convenience whether one chooses to characterize a system by six elastic constants and an angle, or seven elastic constants.

HEXAGONAL All seven hexagonal point groups contain a six-fold rotation axis. The x_3 axis will be taken as the six-fold crystallographic axis. Obviously the hexagonal system possesses a three-fold axis, so the arguments presented for the trigonal point groups 3 and $\bar{3}$ apply. Thus, only the seven independent elastic constants found for that trigonal case need be considered. It is also obvious that the hexagonal system has a two-fold axis pointed along x_3. It follows that any c_{ijkl} for which the index 3 appears an odd number of times must be zero. Thus, $c_{14} = c_{1123}$ and $c_{25} = c_{2213}$ both equal zero. The result is the elastic constant matrix given in Table 2.5, where the x_3 axis points along the six-fold rotation axes, but the orientation of the other axes is arbitrary.

The hexagonal elastic constant matrix has a remarkable property. The matrix is invariant under arbitrary rotations about the six-fold axis, *i.e.*, it is isotropic in the basal plane [6]. This can be demonstrated by taking a two-index elastic constant from the hexagonal matrix of Table 2.5, converting to the four-index form, and then using Equation 2.45 for a rotation about the x_3 axis. For example,

$$\begin{aligned} c_{1111}(\theta) \rightarrow c_{1111}\left(\cos^4\theta + \sin^4\theta\right) + 2\left(c_{1111} - c_{1122}\right)\sin^2\theta\cos^2\theta \\ + 2c_{1122}\sin^2\theta\cos^2\theta. \end{aligned} \tag{2.76}$$

The right-hand side reduces to c_{1111} for any angle θ. Similar results are found for all five independent elastic constants of the hexagonal matrix. The elastic constants are independent of the orientation within in the basal plane.

CUBIC The cubic case is immediately obtained from the orthorhombic one by noting that the x_i axes are all equivalent for cubic symmetry. Thus, $c_{12} = c_{1122} = c_{1133} = c_{13}$, *etc.*

Table 2.6 *Isotropic and icosahedral quasicrystal.*

$$\begin{pmatrix} c_{11} & c_{12} & c_{12} & 0 & 0 & 0 \\ c_{12} & c_{11} & c_{12} & 0 & 0 & 0 \\ c_{12} & c_{12} & c_{11} & 0 & 0 & 0 \\ 0 & 0 & 0 & c_{44} & 0 & 0 \\ 0 & 0 & 0 & 0 & c_{44} & 0 \\ 0 & 0 & 0 & 0 & 0 & c_{44} \end{pmatrix}$$

$$c_{44} = \tfrac{1}{2}\left(c_{11} - c_{12}\right)$$

Two independent elastic constants

Isotropic Materials

Many materials are composed of crystallites that are small on the range scale of interest, and randomly oriented. In such cases, the materials may be considered as isotropic macroscopically. The question to be addressed is, how many independent elastic constants are required to describe the linear elasticity of such a material. Obviously an isotropic material has any of the symmetry elements used, for example, to deduce the elastic constant matrix for cubic symmetry (such as three mutually perpendicular, two-fold rotations axes, and the equivalence of the three x_i axes). Thus, at most three elastic constants are required. Equation 2.45 will be used for an arbitrary rotation of the cubic c_{1111} about the x_3 axis. The transformation gives

$$c_{1111} \rightarrow c_{1111}\left(\cos^4\theta + \sin^4\theta\right) + 2c_{1122}\cos^2\theta\sin^2\theta + 4c_{2323}\cos^2\theta\sin^2\theta. \quad (2.77)$$

Requiring the right-hand side of expression 2.77 to equal c_{1111} gives

$$c_{44} = \frac{1}{2}\left(c_{11} - c_{12}\right). \quad (2.78)$$

Equation 2.78 holds independently of the angle θ; thus, the isotropic case has only two independent elastic constants, as shown in Table 2.6.

Quasicrystals

Quasicrystals comprise a class of materials with long-range aperiodic order, and crystallographically forbidden rotational symmetries. As is well-known, structures with long-range translation symmetries can only possess rotational symmetries of 1,2,3,4,6 order. In contrast, quasicrystals possess other rotational symmetries such as five- and ten-fold. As might be expected, these symmetries impose restrictions on the number of independent elastic constants. The focus here is on icosahedral quasicrystals. These materials are so named, because they have the symmetry elements of an icosahedron.

A convenient picture for the present discussion involves an icosahedron inscribed inside a cube [15]. The corners of the cube are given by $(\pm u, \pm u, \pm u)$, while the vertices of the icosahedron are given by

$$(\pm u, \pm v, 0),\ (0, \pm u, \pm v),\ (\pm v, 0, \pm u), \tag{2.79}$$

where

$$\frac{v}{u} = \tau - 1. \tag{2.80}$$

The parameter τ is known as the golden mean,

$$\tau = \frac{\sqrt{5}+1}{2}. \tag{2.81}$$

The number τ has the unique mathematical properties that $\tau^2 = \tau + 1$ and $1/\tau = \tau - 1$.

The icosahedron inscribed inside the cube is illustrated in Figure 2.16. The x_i axes are located along three mutually perpendicular two-fold rotation axes of the icosahedron, *which coincide with two-fold (also four-fold) rotation axes of the cube*. Similarly, the four, three-fold rotation axes of the cube are aligned with three-fold rotation axes of the icosahedron. (Only one of the three-fold axes is illustrated in Figure 2.16.) By arguments presented in Section 2.3.4 the three two-fold axes are sufficient to restrict the number of independent elastic constants to

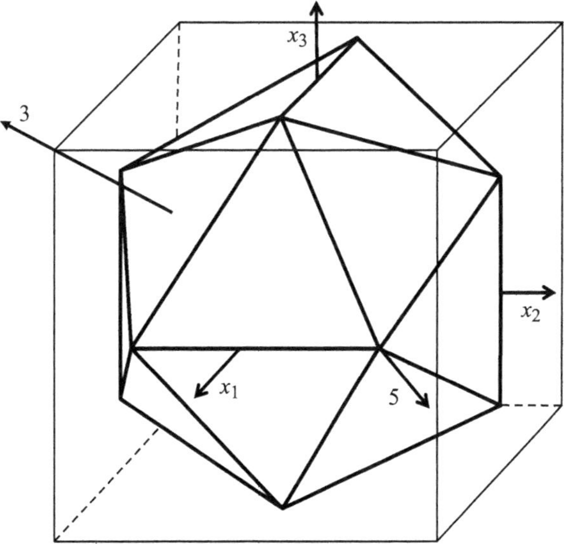

Figure 2.16 Icosahedron inscribed inside a cube. One of the three-fold and one of the five-fold symmetry axes are shown.

those of orthorhombic symmetry. The arguments leading to the cubic case from the orthorhombic case also apply in the present situation, so that the symmetry is further restricted to that of cubic.

An additional restriction on the number of independent elastic constants is obtained by examining the five-fold rotation axes [16], one of which is illustrated in Figure 2.16 (there are a total of six such axes). These axes do not exist in classical crystalline materials. The procedure will be to require that the elastic constant matrix be invariant under a five-fold rotation about one of these axes. The situation is a bit more complicated than the three-fold rotations carried out for the trigonal case. In the present case the x_1 axes will be rotated to the five-fold axis, a rotation by $2\pi/5$ will be performed, and then the inverse of the first rotation will be performed.

By considering the dot product between a vector along the x_1 axis, $(u, 0, 0)$ and a vector along the indicated five-fold axis, $(u, v, 0)$, the angle between these two axes is found to be $\cos^{-1}\theta = \tau/\sqrt{1+\tau^2}$. The set of direction cosines for first rotation is

$$\mathbf{a}_1 = \begin{pmatrix} \frac{\tau}{\sqrt{1+\tau^2}} & \frac{1}{\sqrt{1+\tau^2}} & 0 \\ -\frac{1}{\sqrt{1+\tau^2}} & \frac{\tau}{\sqrt{1+\tau^2}} & 0 \\ 0 & 0 & 1 \end{pmatrix}. \tag{2.82}$$

The next operation is to rotate by $2\pi/5$ about the five-fold axis. Some economy of algebra is obtained by noting that

$$\cos(2\pi/5) = \frac{\sqrt{5}-1}{4} = \frac{\tau - 1}{2}. \tag{2.83}$$

The direction cosines for this second rotation are

$$\mathbf{a}_2 = \begin{pmatrix} 1 & 0 & 0 \\ 0 & \frac{\tau-1}{2} & \frac{\sqrt{\tau+2}}{2} \\ 0 & -\frac{\sqrt{\tau+2}}{2} & \frac{\tau-1}{2} \end{pmatrix}. \tag{2.84}$$

The desired operation is

$$\mathbf{a} = \mathbf{a}_1{}^{-1}\mathbf{a}_2\mathbf{a}_1, \tag{2.85}$$

which after some simplification becomes

$$\mathbf{a} = \frac{1}{2}\begin{pmatrix} \tau & \frac{1}{\tau} & -1 \\ \frac{1}{\tau} & 1 & \tau \\ 1 & -\tau & \frac{1}{\tau} \end{pmatrix}. \tag{2.86}$$

The cubic elastic constants in the coordinate system x_1, x_2, x_3 are $c_{11} = c_{1111}$, $c_{12} = c_{1122}$ and $c_{44} = c_{2323}$, where, as usual, the four index notation is used for transformations between coordinate systems. These elastic constants will be required to be invariant under the five-fold rotation. For example, starting with the cubic symmetry results and applying Equations 2.45 and 2.86 to c_{1122} results in

$$\begin{aligned} c_{1122} \rightarrow \frac{c_{1111}}{16}\left(1+\frac{1}{\tau^2}+\tau^2\right)+\frac{c_{1122}}{16}\left(2+\frac{1}{\tau^2}+\tau^2+\frac{1}{\tau^4}+\tau^4\right) \\ +\frac{c_{2323}}{4}\left(-1+\frac{1}{\tau}-\tau\right). \end{aligned} \tag{2.87}$$

Requiring the right-hand side of Equation 2.87 to equal c_{1122} results in

$$c_{44}=\frac{1}{2}\left(c_{11}-c_{12}\right). \tag{2.88}$$

Subjecting either c_{11} or c_{12} to the same transformation also gives Equation 2.88. The elastic constant matrix for the icosahedral quasicrystal is the same as that for isotropic materials, Table 2.6. This does not mean that the quasicrystal is isotropic for other physical properties, only properties described by a fourth-rank tensor.

There are other types of quasicrystals, some having an axis along which the material is periodic and a perpendicular plane within which the material is quasi-periodic. Various systems have eight-fold, ten-fold, or twelve-fold rotational symmetry. A number of studies have treated these different systems [17, 18, 19, 20, 21].

Anisotropy Due to Materials Processing – Texture

Previous discussions included the single-crystal and isotropic cases, the latter being materials composed of crystallites small compared to the length scale of interest (polycrystals), and randomly oriented. The completely random orientation leads to the isotropic elastic constant matrix, Table 2.6, with just two independent elastic constants. There exists an important intermediate case, where there is a statistically preferred orientation of the crystallites. This preferred orientation of grains is known as texture, and has important effects on many material properties, including elastic constants. Texture can arise due to mechanical processing or heat treatment, and occurs in both engineering and geologic materials.

The arguments leading to the form of the elastic constants matrices depend only on macroscopic symmetry; the underlying atomic nature of the material does not enter. Such is also the case with texture-related elastic constant matrices. There are two important types of texture: rolling, or sheet, texture; and fiber texture. Rolling texture occurs in rolled plates, as the name implies, and in other cases as well. For rolled plates there are three mutually perpendicular directions: the plate rolling direction; the plate transverse direction; and the plate normal direction. In the ideal case this process leads to three mutually perpendicular mirror planes. By the symmetry arguments discussed earlier, these mirror planes lead to an *orthorhombic elastic constant matrix* with nine independent elastic constants. Note that it is not necessary for the crystallites to have orthorhombic symmetry. This microscopic symmetry is averaged away for the length scales much greater that the typical grain size. The underlying material may have any of the various crystalline symmetries.

Fiber texture can occur in drawn wires, where there is orientation of the crystallites along the direction of the wire, but the orientation in the transverse direction is random. Fiber texture also occurs frequently in thin films where one particular cystalline direction grows fastest. For example in face-centered cubic (fcc) materials, the [111] direction frequently grows fastest, leading to a high tendency for crystallites to have their [111] axis perpendicular to the plane of the film, with random orientation in the transverse direction. The symmetry arguments applied to hexagonal symmetry in Section 2.3.4 apply in the present case with the result that the elastic properties of these materials are described by the *hexagonal elastic constant matrix* with five independent elastic constants. Note that the individual crystallites may have any of the crystalline symmetries: cubic, hexagonal, tetragonal, etc.

Texture can be described theoretically in terms of orientation distribution coefficients (ODCs), which are related to the probability to find crystallites oriented in particular directions. Conversely, information about the ODCs can be obtained from measurements of the elastic constants in textured materials [22].

2.3.5 Other Commonly Used Elastic Parameters

Although the elastic constants discussed in the previous section provide a complete characterization of the linear elastic response of materials, other elastic parameters are in common usage. These parameters, in some sense, have simpler physical interpretations than the c_{ijkl}. Among these parameters are the bulk modulus, Young's modulus, Poisson's ratio, and the torsion modulus. These parameters are usually expressed in terms of a combination of the c_{ijkl}. The shear modulus, in most cases of interest, can be expressed in terms of a single one of the c_{ijkl}.

Bulk Modulus

For discussion of the bulk modulus, it will be useful to have an expression for the change in volume produced by a deformation. Consider a cube of material of dimension a on a side. Assume that uniform stresses are applied on each face such that the dimensions of each side increase as $a_1 \to a + \Delta a_1$, $a_2 \to a + \Delta a_2$, and $a_3 \to a + \Delta a_3$. The fractional change in volume is found to be

$$\frac{\Delta V}{V} \approx \frac{\Delta a_1}{a} + \frac{\Delta a_2}{a} + \frac{\Delta a_3}{a} = \epsilon_1 + \epsilon_2 + \epsilon_3, \tag{2.89}$$

where it is assumed that the fractional changes are small, and higher order terms are ignored. It is convenient to work with the compliance matrix, rather than the elastic constant matrix. The two matrices are inverses of each other, and have the same symmetry properties. The compliance matrix gives the strain produced by a given stress. A general form is given in Equation 2.90,

$$\begin{pmatrix} \epsilon_1 \\ \epsilon_2 \\ \epsilon_3 \\ \epsilon_4 \\ \epsilon_5 \\ \epsilon_6 \end{pmatrix} = \begin{pmatrix} s_{11} & s_{12} & s_{13} & s_{14} & s_{15} & s_{16} \\ s_{12} & s_{22} & s_{23} & s_{24} & s_{25} & s_{26} \\ s_{13} & s_{23} & s_{33} & s_{34} & s_{35} & s_{36} \\ s_{14} & s_{24} & s_{34} & s_{44} & s_{45} & s_{46} \\ s_{15} & s_{25} & s_{35} & s_{45} & s_{55} & s_{56} \\ s_{16} & s_{26} & s_{36} & s_{46} & s_{56} & s_{66} \end{pmatrix} \begin{pmatrix} \sigma_1 \\ \sigma_2 \\ \sigma_3 \\ \sigma_4 \\ \sigma_5 \\ \sigma_6 \end{pmatrix}. \quad (2.90)$$

The definition of the bulk modulus is

$$B = -V\frac{dP}{dV} \cong -\frac{\Delta P}{\Delta V/V}. \quad (2.91)$$

For the case of hydrostatic pressure $\sigma_1 = \sigma_2 = \sigma_3 = -\Delta P$, and $\sigma_4 = \sigma_5 = \sigma_6 = 0$. (Pressure, P, is conventionally taken as positive acting inward on the faces of the cube, and σ is taken as positive acting outward.) Combining Equations 2.89, 2.90, and 2.91 gives

$$B = \frac{1}{s_{11} + s_{22} + s_{33} + 2\left(s_{12} + s_{13} + s_{23}\right)}, \quad (2.92)$$

a general expression for the bulk modulus in terms of the compliances. The volume compressibility κ is defined as the inverse of B

$$\kappa = s_{11} + s_{22} + s_{33} + 2\left(s_{12} + s_{13} + s_{23}\right). \quad (2.93)$$

Young's Modulus

Young's modulus, E, is defined as the ratio of a uniaxial stress along a given axis to the resulting longitudinal strain along the same axis. It is a measure of the stiffness of a material, *i.e.*, its resistance to a linear extension or compression along the given axis. Only the single stress is applied. (In general more than one strain will result from the application of this single stress.) Choosing the direction as $\hat{\mathbf{x}}_1$, Equation 2.90 is used again, but in this case all stresses are zero except σ_1,

$$E_1 = \frac{\sigma_1}{\epsilon_1} = \frac{1}{s_{11}}. \quad (2.94)$$

Obviously, if the single stress is along $\hat{\mathbf{x}}_2$ ($\hat{\mathbf{x}}_3$), then Young's modulus is $1/s_{22}$ ($1/s_{33}$). A more general discussion of the directional dependence of Young's modulus will be given later.

Poisson's Ratio

In general a longitudinal stress along a particular axes will give rise to both longitudinal and lateral strains. As an example, if a material is stretched along a given direction it usually contracts in the lateral directions. This effect leads to the definition of Poisson's ratio, ν,

$$\nu_{ij} = -\frac{\epsilon_j}{\epsilon_i} \tag{2.95}$$

Here, ϵ_i denotes the longitudinal strain in response to a longitudinal stress along the x_i axis and ϵ_j the corresponding lateral strain along the x_j axis. Using Equation 2.90 with all stresses equal to zero except σ_1 results in,

$$\nu_{12} = -\frac{s_{12}}{s_{11}} \tag{2.96}$$

and

$$\nu_{13} = -\frac{s_{13}}{s_{11}}. \tag{2.97}$$

Except for high symmetry cases, $\nu_{12} \neq \nu_{13}$. A positive ν_{ij}, the usual case, means that an extension along a given direction leads to a contraction in a perpendicular direction. Negative values for ν_{ij} are unusual, but not unknown [23].

Torsion Modulus

Just as Young's modulus provides a convenient measure of the resistance to extension along a particular direction, the torsion modulus provides a measure of the resistance to twisting about a particular axis. Torsion has many practical applications, *e.g.*, in the transmission of power by a shaft. In addition, the directional dependences of the torsion modulus and Young's modulus provide a convenient means of displaying the anisotropic elastic properties of materials.

The applied stresses in the cases just considered – B, E, and ν – were quite simple. The situation is more complicated for torsion, at least in Cartesian coordinate systems. However, this problem provides a nice example of the solution of the elastic equations, Equation 2.29, for an equilibrium problem subject to boundary conditions.

Torsion is perhaps best visualized as the twisting of a rod due to an applied torque. The problem to be solved is to relate the applied torque to the angle of twist. The situation is illustrated in Figure 2.17. Torques Γ are applied to opposite ends of a cylinder. (A brief discussion as to the details of the application of the torques will be given later.) The shape of the cross-section is arbitrary at this point of the discussion, but it does not vary along the axial direction (cylindrical). The material is assumed to be homogeneous, but may be elastically anisotropic. The straight line P_1P_2 becomes the helical line Q_1Q_2 under the action of the applied

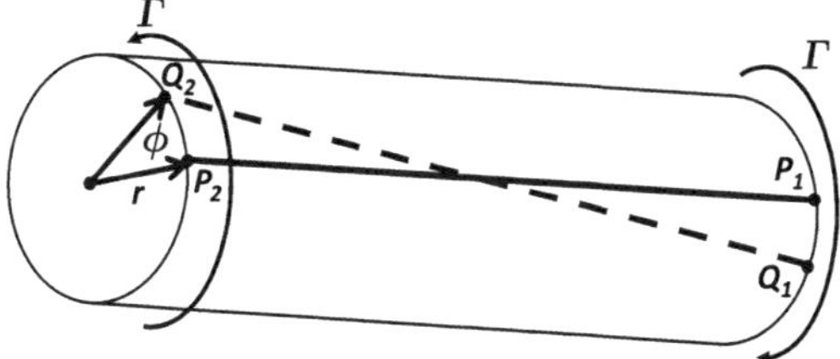

Figure 2.17 Twisting of a cylinder due to torques applied to opposite ends.

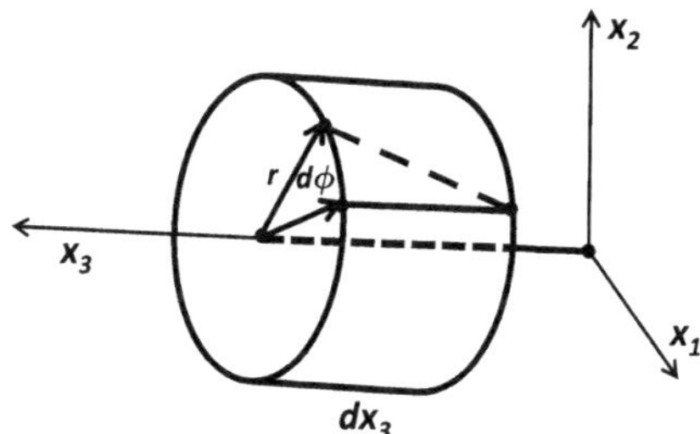

Figure 2.18 Twisting of a thin section of the cylinder.

torques, and the point P_2 at a distance r from the axis of rotation rotates by an angle ϕ to the point Q_2. The magnitude and sign of ϕ depend on the location along the axial direction.

It is instructive to examine a thin section of the cylinder as shown in Figure 2.18. Two neighboring sections a distance dx_3 apart rotate through a relative angle $d\phi$, which is assumed to be proportional to dx_3.

$$d\phi = \tau dx_3, \tag{2.98}$$

where τ is called the twist per unit length.

$$\tau = d\phi/dx_3 \tag{2.99}$$

As there is no variation of properties in the x_3 direction, τ is assumed to be constant.

The approach to the problem is to assume a form for the displacements suggested by Figures 2.17 and 2.18, and then to show that the assumed displacements provide a unique solution to the elastic equations [24]. From Figure 2.18 it is seen that the displacement of a point at the tip of the radius vector $\mathbf{r}$ produced by a rotation of $\Delta\phi$ is [5]

$$\Delta\mathbf{r} = \boldsymbol{\Delta\phi} \times \mathbf{r}, \tag{2.100}$$

where the vector $\boldsymbol{\Delta\phi}$ points in the x_3 direction. The result is

$$\Delta\mathbf{r} = -\Delta\phi x_2 \hat{\mathbf{x}}_1 + \Delta\phi x_1 \hat{\mathbf{x}}_2 \tag{2.101}$$

If ϕ is taken as 0 at $x_3 = 0$ then $\Delta\phi = \tau x_3$, and the components of the displacement in the $x_1 - x_2$ plane are

$$\begin{aligned} u_1 &= -\tau x_2 x_3 \\ u_2 &= \tau x_1 x_3. \end{aligned} \tag{2.102}$$

When the cylinder is twisted, points may in general undergo a displacement along the x_3 axis. Such displacements are, of course, zero for $\tau = 0$, so it is reasonable to assume they are proportional to τ

$$u_3 = \tau\psi(x_1, x_2), \tag{2.103}$$

where $\psi(x_1, x_2)$ is a function to be determined [5]. Equations 2.102 and 2.103 describe a situation where twisting leads to a rotation and warping of a cross section, *i.e.*, a cross section does not, in general, remain plane.

Given the displacements of Equations 2.102 and 2.103, the strains can be calculated using Equation 2.9 and Table 2.2. The only non-zero strains are

$$\begin{aligned} \epsilon_4 &= \tau\left(\frac{\partial\psi}{\partial x_2} + x_1\right) \\ \epsilon_5 &= \tau\left(\frac{\partial\psi}{\partial x_1} - x_2\right). \end{aligned} \tag{2.104}$$

The corresponding stresses are

$$\begin{aligned} \sigma_1 &= c_{14}\epsilon_4 + c_{15}\epsilon_5 \\ \sigma_2 &= c_{24}\epsilon_4 + c_{25}\epsilon_5 \\ \sigma_3 &= c_{34}\epsilon_4 + c_{35}\epsilon_5 \\ \sigma_4 &= c_{44}\epsilon_4 + c_{45}\epsilon_5 \\ \sigma_5 &= c_{45}\epsilon_4 + c_{55}\epsilon_5 \\ \sigma_6 &= c_{46}\epsilon_4 + c_{56}\epsilon_5. \end{aligned} \tag{2.105}$$

The following discussion will be restricted to cases for which there is a two-fold rotation axis along x_3. According to the discussion of 2.3.4 it follows that

$$c_{14} = c_{15} = c_{24} = c_{25} = c_{34} = c_{35} = c_{46} = c_{56} = 0 \tag{2.106}$$

and

$$\begin{aligned} \sigma_4 &= \tau\left(c_{44}\left(\frac{\partial\psi}{\partial x_2} + x_1\right) + c_{45}\left(\frac{\partial\psi}{\partial x_1} - x_2\right)\right) \\ \sigma_5 &= \tau\left(c_{45}\left(\frac{\partial\psi}{\partial x_2} + x_1\right) + c_{55}\left(\frac{\partial\psi}{\partial x_1} - x_2\right)\right) \end{aligned} \tag{2.107}$$

are the only non-zero stresses.

At this point it may be useful to give a brief summary of the torsion discusion to this point, and a preview of the next steps. The physical pictures, Figures 2.17 and 2.18, led intuitively to the assumed form of the displacements for the torsion of a rod, *i.e.*, Equations 2.102 and 2.103. Using these displacements along with the definition of strain, Equation 2.9, and the generalized Hooke's Law, Equation 2.64, gave the corrresponding strains, Equations 2.104, and stresses, Equations 2.107. *To finish the problem, Equations 2.107 will be used along with the elastic equations for equilibrium,*
Equation 2.29, and boundary conditions to solve for the function $\psi(x_1, x_2)$, *which completes the solution of the problem.* This solution will then be applied to a cylinder with an elliptical cross-section and anisotropic elasticity. Simplified cases will also be discussed.

For elastic equilibrium and no body forces, Equation 2.29 becomes

$$\sum_j \frac{\partial \sigma_{ij}}{\partial x_j} = 0 \tag{2.108}$$

for $i = 1, 2, 3$.

Applying Equation 2.108 to the present case where only two stresses are nonzero, and switching back and forth between two-index and one-index notation for the stress gives the following:

$i = 1$

$$\frac{\partial \sigma_5}{\partial x_3} = 0 \tag{2.109}$$

$i = 2$

$$\frac{\partial \sigma_4}{\partial x_3} = 0 \tag{2.110}$$

$i = 3$

$$\frac{\partial \sigma_5}{\partial x_1} + \frac{\partial \sigma_4}{\partial x_2} = 0 \tag{2.111}$$

Equations 2.109 and 2.110 are satisfied automatically as the stresses σ_4 and σ_5 have no dependence on x_3. Combining Equations 2.111 and 2.107 yields

$$c_{55}\frac{\partial^2 \psi}{\partial x_1^2} + 2c_{45}\frac{\partial^2 \psi}{\partial x_1 \partial x_2} + c_{44}\frac{\partial^2 \psi}{\partial x_2^2} = 0 \tag{2.112}$$

Equation 2.112 determines ψ *subject to boundary conditions.*

For boundary conditions it is assumed that the forces acting on the lateral surfaces of the cylinder, called the traction forces, are zero. Equation 2.37 may be used where the forces per unit area, f_i, are assumed to be acting on the lateral surface of the cylinder, and **n** is a normal to that surface. For the present case the stresses are $\sigma_4 = \sigma_{23}$ and $\sigma_5 = \sigma_{13}$ leading to

$$\begin{aligned} f_1 &= \sigma_{13} l_3 \\ f_2 &= \sigma_{23} l_3 \\ f_3 &= \sigma_{13} l_1 + \sigma_{23} l_2. \end{aligned} \tag{2.113}$$

From the geometry, **n** is perpendicular to the x_3 axis, resulting in $l_3 = 0$, so f_1 and $f_2 = 0$ automatically. The third part of Equation 2.113 becomes

$$\sigma_5 l_1 + \sigma_4 l_2 = 0, \tag{2.114}$$

which is to be satisfied on the boundary. Consider an arc length $\mathbf{ds} = dx_1 \hat{\mathbf{x}}_1 + dx_2 \hat{\mathbf{x}}_2$ along the boundary. A normal to that arc length can be written as

$$\mathbf{n} = \frac{dx_2 \hat{\mathbf{x}}_1 - dx_1 \hat{\mathbf{x}}_2}{ds}. \tag{2.115}$$

Then,

$$\begin{aligned} l_1 &= \mathbf{n} \cdot \hat{\mathbf{x}}_1 = \frac{dx_2}{ds} \\ l_2 &= \mathbf{n} \cdot \hat{\mathbf{x}}_2 = -\frac{dx_1}{ds} \end{aligned} \tag{2.116}$$

Combining Equations 2.114 and 2.116 results in

$$\sigma_5 dx_2 - \sigma_4 dx_1 = 0, \tag{2.117}$$

to be satisfied on the boundary. Combining Equations 2.107 and 2.117 leads to

$$\begin{aligned} &\left(c_{55} \left(\frac{\partial \psi}{\partial x_1} - x_2 \right) + c_{45} \left(\frac{\partial \psi}{\partial x_2} + x_1 \right) \right) dx_2 - \\ &\qquad \left(c_{44} \left(\frac{\partial \psi}{\partial x_2} + x_1 \right) + c_{45} \left(\frac{\partial \psi}{\partial x_1} - x_2 \right) \right) dx_1 = 0 \end{aligned} \tag{2.118}$$

A solution to the torsion problem requires that Equations 2.112 and 2.118 be satisfied simultaneously [24].

Elliptical Cross-Section Two simplifications will be made at this point. The previous restrictions on the elastic constants, Equation 2.106, excluded the triclinic and trigonal cases (Tables 2.4 and 2.5). For the following discussions, c_{45} will be taken as zero, which excludes the monoclinic case as well. The second simplification is the restriction to an elliptical cross-section. The x_3 axis coincides with the elastic symmetry axis and the symmetry axis of the cylinder. Equations 2.112 and 2.118 become

$$c_{55}\frac{\partial^2\psi}{\partial x_1^2} + c_{44}\frac{\partial^2\psi}{\partial x_2^2} = 0 \tag{2.119}$$

$$c_{55}\left(\frac{\partial\psi}{\partial x_1} - x_2\right)dx_2 = c_{44}\left(\frac{\partial\psi}{\partial x_2} + x_1\right)dx_1 \tag{2.120}$$

Equation 2.119 is to hold throughout the rod and Equation 2.120 is to hold on the boundary. A trial solution will be assumed [24],

$$\psi(x_1, x_2) = Ax_1x_2 \tag{2.121}$$

where A is to be determined. Equation 2.119 is obviously satisfied. Equation 2.120 becomes

$$c_{44}(A+1)x_1dx_1 = c_{55}(A-1)x_2dx_2 \tag{2.122}$$

Integrating results in

$$c_{44}(1+A)x_1^2 + c_{55}(1-A)x_2^2 = C \tag{2.123}$$

where C is a constant. Comparing Equation 2.123 with the equation for an ellipse in standard form,

$$\frac{x_1^2}{a^2} + \frac{x_2^2}{b^2} = 1 \tag{2.124}$$

gives

$$A = \frac{b^2c_{55} - a^2c_{44}}{a^2c_{44} + b^2c_{55}} \tag{2.125}$$

and

$$C = \frac{2a^2b^2c_{44}c_{55}}{a^2c_{44} + b^2c_{55}}, \tag{2.126}$$

where a and b are the semi-major and semi-minor axes of the ellipse. Thus, Equation 2.123 which holds on the boundary (assumed elliptical), indeed describes an ellipse where the constants A and C are related to the semi-axes of the ellipse by Equations 2.125 and 2.126. Using Equation 2.107 with $c_{45} = 0$, and Equations 2.121, it is possible to obtain the stresses in terms of the elastic constants and the parameters of the elliptical cross-section

$$\begin{aligned}\sigma_4 &= \tau c_{44}(A+1)x_1\\ \sigma_5 &= \tau c_{55}(A-1)x_2.\end{aligned} \tag{2.127}$$

Equations 2.127 essentially constitute a solution to the problem; we now have the stresses corresponding to the displacements, Equations 2.102 and 2.103, and associated strains, Equations 2.104. The function ψ is given by Equations 2.121 and 2.125. The situation is restricted to an elliptical cross-section, and excludes triclinic, trigonal, and monoclinic crystal symmetries.

Although a solution has been obtained, it is more useful in the present case to present the solution in terms of torques rather than stresses. The procedure will be to first find the forces on the ends of the rod of Figure 2.18 and then find the torques. Equation 2.37 will be used where, for the present application, $l_1 = l_2 = 0$ and $l_3 = \pm 1$ where the + and – signs are for the left and right ends of the rod in Figure 2.18 respectively. The results for the left end of the rod are

$$\begin{aligned} f_1 &= \sigma_5 \\ f_2 &= \sigma_4. \end{aligned} \tag{2.128}$$

As the f_i are forces per unit area, they must be integrated over the ends of the rod. Because the forces are odd functions of the x_i – Equations 2.127 and 2.128 – the integration will give zero for the net forces on the ends of the rod. However, these forces do produce a torque on the end face as illustrated in Figure 2.19. Considering the forces acting at the point with coordinates (x_1, x_2), the torque is found to be

$$\Gamma = \int\int (f_2 x_1 - f_1 x_2) dx_1 dx_2. \tag{2.129}$$

Equation 2.129 is true quite generally, subject to the restriction to a two-fold symmetry axis, Equation 2.106.

Restricting the calculation to the elliptical cross-section, Equations 2.127, 2.128, and 2.129 result in

$$\Gamma = \tau\,(c_{44}(A+1)I_2 - c_{55}(A-1)I_1) \tag{2.130}$$

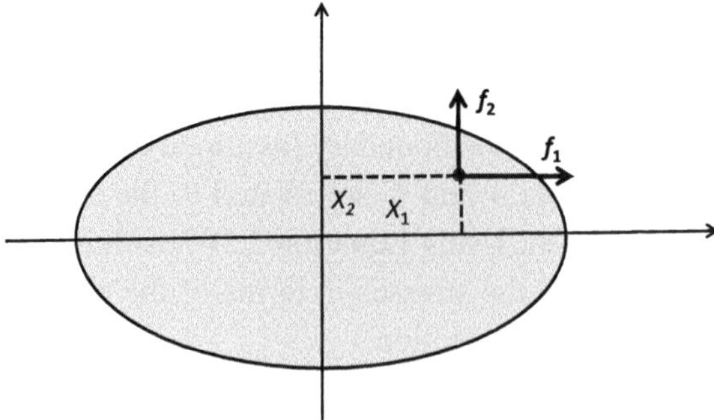

Figure 2.19 Forces on the end face of an elliptical rod.

where

$$I_1 = \int\int x_2^2 dx_1 dx_2 = \frac{\pi ab^3}{4}$$
$$I_2 = \int\int x_1^2 dx_1 dx_2 = \frac{\pi a^3 b}{4}. \quad (2.131)$$

Using $1/c_{44} = s_{44}$ and $1/c_{55} = s_{55}$, along with Equations 2.125, 2.130, and 2.131, the expression for Γ can be simplified considerably,

$$\Gamma = \frac{\pi a^3 b^3}{s_{44} b^2 + s_{55} a^2} \tau. \quad (2.132)$$

Equation 2.132 relates the twist, τ, to the applied torque for an elastically anisotropic rod with an ellipitcal cross-section. The symmetry axis of the rod is along the x_3 direction. *The factor relating the twist to the torque is commonly called the torsional rigidity* [25, 26],

$$\text{Torsional rigidity} = \frac{\pi a^3 b^3}{s_{44} b^2 + s_{55} a^2}. \quad (2.133)$$

Note that the torsional rigidity depends not only on the material properties, s_{44} and s_{55}, but also on the structure, in this case the semi-axes of the ellipse, a and b.

From Equations 2.103, 2.121, and 2.125 the warping is given by

$$u_3 = \tau \left(\frac{b^2 c_{55} - a^2 c_{44}}{a^2 c_{44} + b^2 c_{55}} \right) x_1 x_2. \quad (2.134)$$

Next, the special case of a circular cross-section is considered, in which case $a = b = R$, and

$$\text{Torsional rigidity} = \frac{\pi R^4}{s_{44} + s_{55}} = \frac{\pi R^4 G_3}{2}, \quad (2.135)$$

where

$$G_3 = \frac{2}{s_{44} + s_{55}} = \frac{1}{2(s_{2323} + s_{1313})} \quad (2.136)$$

is the *torsional modulus* for rotation about the x_3 axis. It is a purely material-dependent parameter characterizing the resistance to twisting (the higher G_3 is, the more difficult it is to twist the material). For rotation about the x_1 or x_2 axes the results are, respectively,

$$G_1 = \frac{2}{s_{55} + s_{66}} = \frac{1}{2(s_{1313} + s_{1212})}$$
$$G_2 = \frac{2}{s_{44} + s_{66}} = \frac{1}{2(s_{2323} + s_{1212})}. \quad (2.137)$$

For Equation 2.132 to be an exact solution, the torque must be produced by forces as given by Equations 2.127 and 2.128. However, the solution is not, in practice, so strictly limited. For a long, twisted bar, the stresses depend almost entirely on the magnitude of the applied torque, and are almost independent of the way in which the forces are distributed over the ends [25].

Shear Modulus

The shear modulus is the ratio of the shear stress to the shear strain. This term, usually denoted by G, is most commonly used to describe isotropic materials. From Table 2.6 it is seen that $G = c_{44}$ for isotropic materials. Because the materials are isotropic, a shear stress on any face is related to the resulting shear strain by G. For crystalline materials, a shear modulus can be defined for any orientation, but the term is less-commonly used for crystalline materials. Note the (unfortunate) similar notation for shear modulus and torsion modulus.

Forms of B, E, ν, and G_i for Various Symmetries

Next, the detailed forms for B, E, ν, and G_i will be developed for several of the different elastic constants matrices of Tables 2.4, 2.5, and 2.6.

Cubic symmetry For cubic symmetry Equation 2.92 reduces to

$$B_{cubic} = \frac{1}{3(s_{11} + 2s_{12})}. \tag{2.138}$$

This equation can be expressed in terms of the elastic constants, rather than the compliances, by inverting the cubic elastic constant matrix of Table 2.4,

$$s_{cubic} = \begin{pmatrix} \frac{c_{11}+c_{12}}{C} & -\frac{c_{12}}{C} & -\frac{c_{12}}{C} & 0 & 0 & 0 \\ -\frac{c_{12}}{C} & \frac{c_{11}+c_{12}}{C} & -\frac{c_{12}}{C} & 0 & 0 & 0 \\ -\frac{c_{12}}{C} & -\frac{c_{12}}{C} & \frac{c_{11}+c_{12}}{C} & 0 & 0 & 0 \\ 0 & 0 & 0 & \frac{1}{c_{44}} & 0 & 0 \\ 0 & 0 & 0 & 0 & \frac{1}{c_{44}} & 0 \\ 0 & 0 & 0 & 0 & 0 & \frac{1}{c_{44}} \end{pmatrix}. \tag{2.139}$$

where

$$C = c_{11}^2 + c_{11}c_{12} - 2c_{12}^2 \tag{2.140}$$

Combining the three previous equations yields

$$B_{cubic} = \frac{c_{11} + 2c_{12}}{3}. \tag{2.141}$$

Young's modulus is easily expressed in terms of the elastic constants by using Equation 2.94,

$$E_1 = \frac{1}{s_{11}} = \frac{c_{11}^2 + c_{11}c_{12} - 2c_{12}^2}{c_{11} + c_{12}}. \tag{2.142}$$

The subscript 1 indicates that the equation applies for a tensile stress along the x_1 axis. Of course, for cubic symmetry the same result is obtained for stresses along the x_2 and x_3 directions.

In a similar manner Poisson's ratio, Equation 2.95, is expressed in terms of the elastic constants using the compliance matrix,

$$\nu_{12} = -\frac{s_{12}}{s_{11}} = \frac{c_{12}}{c_{11} + c_{12}}. \tag{2.143}$$

Recalling 2.95, it is seen that $\nu_{12} > 0$ means that an extension along the x_1 axis is accompanied by a contraction along the x_2 axis. For cubic symmetry, all the ν_{ij} are equal for $i, j = 1, 2, 3$, and $i \neq j$

For the torsional modulus the result is $G_1 = G_2 = G_3 = c_{44}$.

Isotropy The isotropic case is easily handled by noting that setting $c_{44} = \frac{1}{2}(c_{11} - c_{12})$ in the cubic case gives the isotropic case. With some simplification the results are

$$\begin{aligned} B &= \frac{3c_{11} - 4c_{44}}{3} \\ E &= c_{44}\frac{(3c_{11} - 4c_{44})}{c_{11} - c_{44}} \\ \nu &= \frac{c_{11} - 2c_{44}}{2(c_{11} - c_{44})}. \end{aligned} \tag{2.144}$$

The elastic constant c_{44}, the shear modulus, describes the resistance to shear, and is often denoted by G. It is also called the rigidity modulus. (The torsional modulus, above, is not always equal to G, *e.g.*, see Section 2.3.5) There are only two independent moduli for the isotropic case, and these are taken to be c_{11} and c_{44} in the above expressions. In some cases it is convenient to take B and G as the independent moduli. In this case E and ν are given by

$$E = \frac{9GB}{3B + G} \tag{2.145}$$

and

$$\nu = \frac{3B - 2G}{2(3B + G)}. \tag{2.146}$$

For the torsional modulus, $G_1 = G_2 = G_3 = c_{44}$.

Both B and G must be ≥ 0 for stability. Applying these limits in turn to Equation 2.146 gives

$$-1 \leq \nu \leq 1/2 \tag{2.147}$$

The upper limit corresponds to an incompressible material, $B \rightarrow \infty$. It is important to remember that the limits on ν were derived for an isotropic material in the linear elastic regime. Also, these are *limits*. It is not a guarantee that any material will reach these limits.

Hexagonal symmetry For a slightly more complicated application of the previous ideas, hexagonal symmetry is considered. The compliance matrix is found by taking the inverse of the hexagonal elastic constant matrix Table 2.5,

$$s_{hexagonal} = \frac{1}{\alpha}\begin{pmatrix} c_{11}c_{33}-c_{13}^2 & c_{13}^2-c_{12}c_{33} & -c_{13}(c_{11}-c_{12}) & 0 & 0 & 0 \\ c_{13}^2-c_{12}c_{33} & c_{11}c_{33}-c_{13}^2 & -c_{13}(c_{11}-c_{12}) & 0 & 0 & 0 \\ -c_{13}(c_{11}-c_{12}) & -c_{13}(c_{11}-c_{12}) & c_{11}^2-c_{12}^2 & 0 & 0 & 0 \\ 0 & 0 & 0 & \frac{\alpha}{c_{44}} & 0 & 0 \\ 0 & 0 & 0 & 0 & \frac{\alpha}{c_{44}} & 0 \\ 0 & 0 & 0 & 0 & 0 & \frac{2\alpha}{c_{11}-c_{12}} \end{pmatrix}. \tag{2.148}$$

where

$$\alpha = (c_{11}-c_{12})\left[c_{33}(c_{11}+c_{12}) - 2c_{13}^2\right]. \tag{2.149}$$

Using Equation 2.92 leads to

$$B_{hexagonal} = \frac{c_{12}c_{33}+c_{11}c_{33}-2c_{13}^2}{c_{11}+c_{12}+2c_{33}-4c_{13}} \tag{2.150}$$

for the bulk modulus of an hexagonal crystal.

Using the compliance matrix and Equation 2.94 gives

$$E_1 = \frac{(c_{11}-c_{12})\left[c_{33}\,(c_{11}+c_{12}) - 2c_{13}^2\right]}{c_{11}c_{33}-c_{13}^2}, \tag{2.151}$$

Young's modulus for the x_1 axis. A similar procedure for $E_3 = 1/s_{33}$ yields,

$$E_3 = \frac{c_{33}(c_{11}+c_{12}) - 2c_{13}^2}{c_{11}+c_{12}}, \tag{2.152}$$

Young's modulus for the x_3 axis.

Equations 2.95, 2.96, 2.97, and 2.148 easily give expressions for various Poisson's ratios. The ratio

$$\nu_{12} = \frac{c_{12}c_{33} - c_{13}^2}{c_{11}c_{33} - c_{13}^2} \tag{2.153}$$

gives the lateral strain along the x_2 axis in response to a strain along the x_1 axis, while

$$\nu_{13} = \frac{c_{13}(c_{11} - c_{12})}{c_{11}c_{33} - c_{13}^2} \tag{2.154}$$

gives the lateral strain along the x_3 axis in response to a strain along the x_1 axis. Other expressions for Poisson's ratio are easily obtained for a uniaxial strain along the x_3 axis. Note that $\nu_{13} \neq \nu_{31}$.

For the torsional modulus, using Equations 2.136, 2.137, and 2.148 yields

$$G_1 = G_2 = \frac{2c_{44}(c_{11} - c_{12})}{c_{11} - c_{12} + 2c_{44}} \tag{2.155}$$

and

$$G_3 = c_{44}. \tag{2.156}$$

While the hexagonal results were derived with single crystals of hexagonal symmetry in mind, they are of more general validity. They are valid for any material which is transversely isotropic, *i.e.*, the material properties are the same in all directions in a plane perpendicular to the axis of symmetry (usually taken to be the x_3 axis). Such materials are sometimes called polar anisotropic. Examples occur in materials science and geophysics, *e.g.*, fiber bundles and thin films, as well as naturally occuring rock formations. Of course, here isotropic is taken to mean that any spatial variations occur on a scale small compared to the appropriate length scale, usually the ultrasonic wavelength.

2.3.6 Polycrystalline Elastic Constants from Single-Crystal Data

The most accurate elastic constants measurements are normally those made on well-characterized single crystals; however, polycrystalline materials are overwhelmingly in wider use than single-crystal materials. Measurements on polycrystalline materials are more likely to be affected by poorly characterized non-uniformity, anisotropic effects, voids, *etc.*, than results from single-crystal materials. Thus, it is desirable to calculate the elastic properties of polycrystalline materials from the single-crystal data to provide ideal polycrystalline elastic constant results. This problem is a difficult one and has a long history. The various averaging results usually provide upper and lower bounds for the polycrystalline moduli. The problem will not be treated here as there are good summaries available in the literature [27, 28].

2.3.7 Representational Surfaces

The tensorial nature of elasticity makes it a challenge to visualize the elastic properties of materials. Graphical methods can help. For anisotropic cases, no single graph, or even two graphs, can relay all the information, but two types of three-dimensional representations have been found useful: Young's modulus and the torsional modulus. In these plots, a radius vector is taken as proportional to the relevant modulus in that direction. Equations for the directional dependences of the various compliances associated with Young's modulus and the torsional modulus have been given in various places [4, 29]. These expressions are rather lengthy, especially for the lower symmetry cases. For the present numerical examples, Equation 2.45 will be used, which applies for the s_{ijkl} as well as for the c_{ijkl}. It is crucial to remember that for Equation 2.45 to apply, the four-index notation must be used.

To use Equation 2.45 it is necessary to define an x_i' coordinate system, and the set of direction cosines between the new and old coordinate systems. The x_i' system is defined as follows: $\hat{\mathbf{x}}_1'$ lies along the direction given by θ and ϕ, the usual angles in spherical coordinates; $\hat{\mathbf{x}}_2'$ axis lies in the $x_1 - x_2$ plane, perpendicular to $\hat{\mathbf{x}}_1'$; and $\hat{\mathbf{x}}_3'$ is found by taking the cross product of $\hat{\mathbf{x}}_1'$ with $\hat{\mathbf{x}}_2'$ [30]. The resulting set of direction cosines can be written as a three by three matrix,

$$\mathbf{a} = \begin{pmatrix} \sin\theta\cos\phi & \sin\theta\sin\phi & \cos\theta \\ -\sin\phi & \cos\phi & 0 \\ -\cos\theta\cos\phi & -\cos\theta\sin\phi & \sin\theta \end{pmatrix}. \tag{2.157}$$

(Note: the directions of $\hat{\mathbf{x}}_2'$, and that of $\hat{\mathbf{x}}_3'$ which follows, affect the intermediate steps of the calculations, but not the final results.)

For what follows, the $\hat{\mathbf{x}}_1'$ direction is taken as the direction for the measurement of Young's modulus and the torsion modulus in the x_i' coordinate system, *i.e.*, an extension along $\hat{\mathbf{x}}_1'$ for Young's modulus and a twisting motion about $\hat{\mathbf{x}}_1'$ for the torsion modulus. The angles θ and ϕ will be used to indicate these quantities, *e.g.*, $E(\theta, \phi)$ means Young's modulus along the $\hat{\mathbf{x}}_1'$ direction,

$$E(\theta, \phi) = \frac{1}{s_{1111}(\theta, \phi)} \tag{2.158}$$

and

$$G(\theta, \phi) = \frac{1}{2(s_{1313}(\theta, \phi) + s_{1212}(\theta, \phi))}. \tag{2.159}$$

Note that G_1 from Equation 2.137 is used for $G(\theta, \phi)$ because $\hat{\mathbf{x}}_1'$ is taken as the unique axis in the x_i' coordinate system.

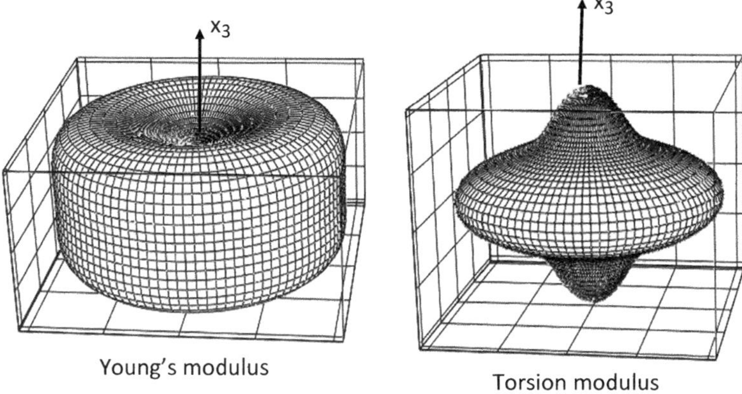

Figure 2.20 Young's and torsion moduli for hexagonal zinc.

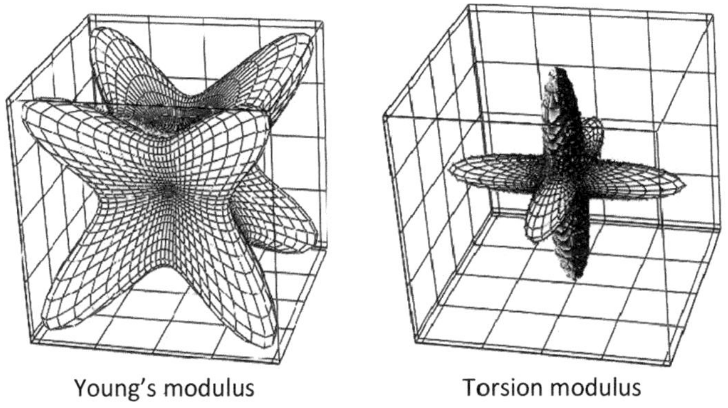

Figure 2.21 Young's and torsion moduli for body-centered cubic potassium.

It simply remains to apply Equations 2.45 (with s_{ijkl} in place of c_{ijkl}), 2.157, 2.158, and 2.159 over a sufficient range and number of angles (θ and ϕ) and plot the results in 3D. An example is given in Figure 2.20. Zinc has the hexagonal structure, and as demonstrated earlier, Equation 2.76, the elastic properties are isotropic in the basal plane. This feature is illustrated nicely in Figure 2.20. The scales are not the same for the two moduli. To give a feeling for the results, along the x_3 axis $E = 35.4$ GPa and $G = 38.3$ GPa, while in the basal plane $E = 118.9$ GPa and $G = 48.2$ GPa.

For another example, consider Figure 2.21 for bcc potassium. The boxes enclosing the figures are aligned along the x_1, x_2, x_3 axes. The general features illustrated in Figure 2.21 seem to be characteristic of the bcc alkali metals and are a result of the structure and the interatomic bonding. The scales are not the same for the two plots in the figure. Young's modulus is a maximum in [111] directions at 4.7 GPa, while the torsion modulus is maximum in [100] type directions at 1.9 GPa.

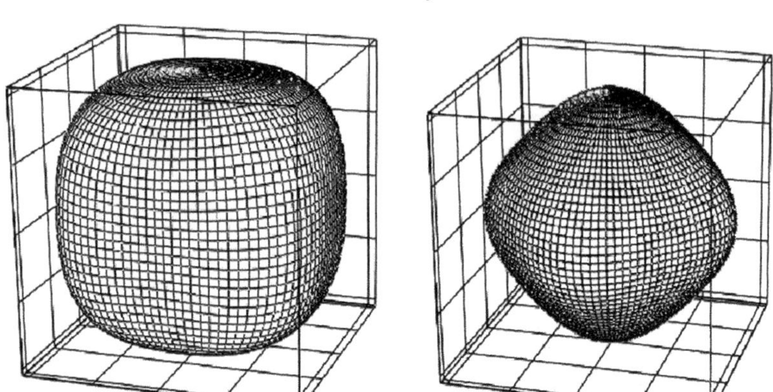

Figure 2.22 Young's and torsion moduli for face-centered cubic aluminum.

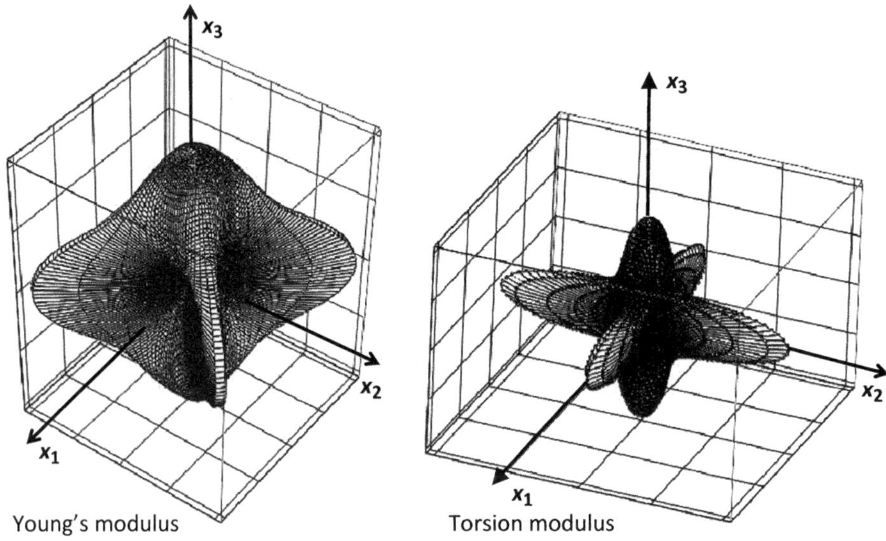

Figure 2.23 Young's and torsion moduli for tetragonal TeO_2.

Figure 2.22 presents yet another example of the Young's and torsion moduli for a common material, in this case fcc aluminum. As in Figure 2.21 the boxes enclosing the figures are aligned with the crystalline cubic axes. The scales are different for the two moduli: Young's modulus along the [111] directions is 75.4 GPa, and the torsion modulus along the [100] directions is 28.3 GPa.

The preceding three Figures demonstrate the wide range of elastic properties exhibited by three common materials. For example, potassium clearly has a much more anisotropic behavior than aluminum. A common measure of anisotropy in cubic materials is the parameter A defined as

$$A = \frac{2c_{44}}{c_{11} - c_{12}} \tag{2.160}$$

Computing A for these two materials gives: Al, $A = 1.23$; K, $A = 6.7$. These numbers are in agreement with Figures 2.21 and 2.22, although clearly the plots give much more information.

While the three previous examples dealt with well-known materials, the final example concerns TeO_2, a less common material, but one which measurements have shown to be exceptionally anisotropic. The unusual properties have been attributed to the crystal structure and interatomic bonding [31]. The material is presented here, however, only as an example of strong anisotropy. Young's modulus reaches a maximum of 117.6 GPa in the [110] directions. The ratio of $E_{[110]}/E_{[100]} = 14.0$. Similarly, G reaches a maximum of 39.2 GPa in the [100] directions, and $G_{[100]}/G_{[110]} = 9.7$.

3

Acoustic Waves in Solids

The interactions among the atoms in a solid enable the propagation of waves. In contrast to electromagnetic waves, which involve oscillating electric and magnetic fields, the waves of interest here involve mechanical movements of the atoms or molecules. Also in contrast to electromagnetic waves, which are transverse in nature, the mechanical waves in a solid have both transverse and longitudinal character. Liquids and gases also support mechanical waves, but only the longitudinal polarizations are possible due to the absence of shear moduli in those states of matter. Such waves in gases (air) are, of course, sound waves. This term is extended to solids (for both longitudinal and transverse polarizations) where the waves are called "infrasound," "sound," or "ultrasound" depending on whether the frequency is respectively below, in, or above the audible range. In common usage the term "acoustics" seems to refer to all types of mechanical waves in solids, liquids, and gases, and that is the sense in which the term is used in the present work.

The treatment of acoustic waves in solids falls naturally into two regimes, depending on wavelength. If the wavelength is much greater than the interatomic spacing, the material is treated as a continuum and classical elasticity applies. For shorter wavelengths the discrete nature of the lattice must be taken into account and the approach of lattice dynamics is used.

3.1 Acoustic Waves in the Classical Elasticity Limit

Acoustic waves described by classical elasticity find many uses. Just to mention a few, these uses involve basic problems in condensed matter physics, material inspections, device operations, and seismology. The basic equation of motion was developed previously, Equation 2.29, but will be repeated here for convenience,

$$\sum_{j=1}^{3} \frac{\partial \sigma_{ij}}{\partial x_j} = \rho \frac{\partial^2 u_i}{\partial t^2}. \tag{3.1}$$

Using Equations 2.9, 2.42, and 2.46, and making an exchange of dummy indices within a sum produces,

$$\sum_j \frac{\partial \sigma_{ij}}{\partial x_j} = \sum_{jkl} c_{ijkl} \frac{\partial^2 u_k}{\partial x_j \partial x_l}, \tag{3.2}$$

where, as usual, i, j, k run from 1 to 3. Combining the previous equations gives,

$$\sum_{jkl} c_{ijkl} \frac{\partial^2 u_k}{\partial x_j \partial x_l} = \rho \frac{\partial^2 u_i}{\partial t^2} \tag{3.3}$$

for the new equation of motion in terms of the displacements u_i.

The next step is to assume plane wave solutions

$$u_i = u_i^o \exp\left(i(\vec{K} \cdot \vec{r} - \omega t)\right), \tag{3.4}$$

where $\vec{K}$ is the wave vector ($K = 2\pi/\lambda$, with λ being the wavelength of a wave traveling in the direction of $\vec{K}$), ω is the angular frequency of the wave, and the subscript i takes the values 1, 2, or 3, corresponding to displacements along the x_1, x_2 or x_3 axes.[1] Note that $\vec{K} \cdot \vec{r} = \sum_l K_l x_l$. Substituting Equation 3.4 in Equation 3.3, taking derivatives and dividing out the common exponential factor results in

$$\sum_{jkl} (c_{ijkl} K_j K_l - \rho \omega^2 \delta_{ik}) u_k^o = 0. \tag{3.5}$$

Now define $K_j = K \cos(\theta_j) = K n_j$ *,where n_j is the cosine of the angle between $\vec{K}$ and the x_j axis*; n_j will be used extensively in what follows.[2] Substituting this definition of K_j into the above equation and dividing by K^2 gives

$$\sum_{jkl} (c_{ijkl} n_j n_l - \rho v^2 \delta_{ik}) u_k^o = 0, \tag{3.6}$$

with

$$v = \omega / K \tag{3.7}$$

being the *phase velocity of the wave*. The physical interpretation of the phase velocity will be discussed in Section 3.1.4. The wave equation for plane wave solutions has become a set of linear equations in the displacement amplitudes u_k^o. One more definition achieves the desired result

$$\Gamma_{ik} = \sum_{jl} n_j n_l c_{ijkl}. \tag{3.8}$$

[1] Notice the double use of i as an index and also as $\sqrt{-1}$.

[2] The set (n_1, n_2, n_3) gives the direction of $\vec{K}$, the propagation direction. It is useful to remember that $n_1^2 + n_2^2 + n_3^2 = 1$.

Then, Equation 3.6 becomes

$$\sum_k (\Gamma_{ik} - \rho v^2 \delta_{ik}) u_k^o = 0. \tag{3.9}$$

Equation 3.9 is actually a set of three equations for $i = 1, 2, 3$, known as the Christoffel equations [32]. It seems worthwhile to write Equations 3.9 more explicitly in matrix form.[3]

$$\begin{pmatrix} \Gamma_{11} - \rho v^2 & \Gamma_{12} & \Gamma_{13} \\ \Gamma_{12} & \Gamma_{22} - \rho v^2 & \Gamma_{23} \\ \Gamma_{13} & \Gamma_{23} & \Gamma_{33} - \rho v^2 \end{pmatrix} \begin{pmatrix} u_1^o \\ u_2^o \\ u_3^o \end{pmatrix} = 0. \tag{3.10}$$

As usual, the condition for a nontrivial solution of Equations 3.10 is that the determinant of the coefficients of the u_i^o be zero,

$$\begin{vmatrix} \Gamma_{11} - \rho v^2 & \Gamma_{12} & \Gamma_{13} \\ \Gamma_{12} & \Gamma_{22} - \rho v^2 & \Gamma_{23} \\ \Gamma_{13} & \Gamma_{23} & \Gamma_{33} - \rho v^2 \end{vmatrix} = 0. \tag{3.11}$$

Equations 3.10 and 3.11 are to be solved for a particular wave vector $\vec{K}$ to find the eigenvalues and eigenvectors for that propagation direction. Solving Equation 3.11 will give three eigenvalues (values of ρv^2) giving three phase velocities for each propagation direction. Using *each* separate value of ρv^2 in Equation 3.10 will give an associated eigenvector ($\vec{u}^o = \sum_{i=1}^{3} u_i^o \hat{\mathrm{x}}_i$). Because multiplying $\vec{u}$ by any scalar will not change the solution of Equation 3.10, the interest is in the *direction* of $\vec{u}$. Solutions of Equation 3.10 and 3.11 thus give three phase velocities and associated directions of $\vec{u}$ for a particular propagation direction.

Because the Christoffel matrix is real and symmetric, the eigenvalues are real. Also, quite importantly, the three solutions for $\vec{u}$ associated with the three eigenvalues are mutually orthogonal. It sometimes happens that two eigenvalues, call them 2 and 3, have the same value (degenerate). In that case the solutions only require that $\vec{u}_2$ and $\vec{u}_3$ lie in a plane perpendicular to $\vec{u}_1$, But, any linear combination of $\vec{u}_2$ and $\vec{u}_3$ is also a solution and it is easy to choose two such solutions that are orthogonal to each other. (They are automatically orthogonal to $\vec{u}_1$.)

Next, solutions of the Christoffel equations will be investigated for various crystalline symmetries.

3.1.1 Cubic Symmetry Solutions

Taking into account the cubic elastic constant matrix of Table 2.5, the Christoffel equation for cubic symmetry becomes [14]

[3] The Γ matrix is symmetric as can be seen by switching the dummy indices of summation in Equation 3.8 and using Equation 2.62 or Equation 2.63.

$$\begin{pmatrix} n_1^2c_{11} + (n_2^2 + n_3^2)c_{44} - \rho v^2 & n_1n_2(c_{12} + c_{44}) & n_1n_3(c_{12} + c_{44}) \\ n_1n_2(c_{12} + c_{44}) & n_2^2c_{11} + (n_1^2 + n_3^2)c_{44} - \rho v^2 & n_2n_3(c_{12} + c_{44}) \\ n_1n_3(c_{12} + c_{44}) & n_2n_3(c_{12} + c_{44}) & n_3^2c_{11} + (n_1^2 + n_2^2)c_{44} - \rho v^2 \end{pmatrix} \begin{pmatrix} u_1^o \\ u_2^o \\ u_3^o \end{pmatrix} = 0. \tag{3.12}$$

[100] Direction

As a simple example the propagation of waves along the [100] direction, *i.e.* along x_1, is considered. In this case $n_1 = 1$ and $n_2 = n_3 = 0$. The determinant takes the simple form,

$$\begin{vmatrix} c_{11} - \rho v^2 & 0 & 0 \\ 0 & c_{44} - \rho v^2 & 0 \\ 0 & 0 & c_{44} - \rho v^2 \end{vmatrix} = 0, \tag{3.13}$$

with the solutions $c_{11} = \rho v^2$, one solution; and $c_{44} = \rho v^2$, two solutions. The displacement amplitudes can be found by using these solutions in Equation 3.10, which in the present case takes the simple form,

$$\begin{pmatrix} c_{11} - \rho v^2 & 0 & 0 \\ 0 & c_{44} - \rho v^2 & 0 \\ 0 & 0 & c_{44} - \rho v^2 \end{pmatrix} \begin{pmatrix} u_1^o \\ u_2^o \\ u_3^o \end{pmatrix} = 0. \tag{3.14}$$

First eigenvalue, $c_{11} = \rho v_l^2$. (As will be discussed shortly, this solution represents a longitudinal wave, hence the subscript on v). Substituting this value into Equation 3.14 gives three equations

$$\begin{aligned} 0u_1^o &= 0 \\ (c_{44} - c_{11})\, u_2^o &= 0\ , \\ (c_{44} - c_{11})\, u_3^o &= 0 \end{aligned} \tag{3.15}$$

resulting in $u_2^o = u_3^o = 0$ with no restriction on u_1^o. The direction of particle motion is along the x_1 axis as is the direction of travel of the wave, so this is a longitudinal wave with speed $v_l = \sqrt{c_{11}/\rho}$.

Second and third eigenvalues, $c_{44} = \rho v^2$**.** Substituting this solution into Equation 3.14 gives $u_1^o = 0$ with no restrictions on u_2^o or u_3^o. It is common, but not necessary, to write the solutions as

$$\vec{u}_2 = u_2^o \begin{pmatrix} 0 \\ 1 \\ 0 \end{pmatrix}, \tag{3.16}$$

and

$$\vec{u}_3 = u_3^o \begin{pmatrix} 0 \\ 0 \\ 1 \end{pmatrix}. \tag{3.17}$$

These displacements are perpendicular to the direction of travel of the wave, and thus represent transverse waves polarized along the x_2 or x_3 axis with speed

$v_t = \sqrt{c_{44}/\rho}$. The overall solution is a pure mode where the particle motion is strictly longitudinal or strictly transverse, and the three $\vec{u}_i$ are mutually perpendicular.

Because of the doubly degenerate eigenvalues, any linear combination of Equations 3.16 and 3.17 is also a transverse wave solution with the same speed. It is straightforward to find two such solutions which are also mutually perpendicular. Therefore any wave polarized perpendicular to the [100] direction, with travel direction along the [100] direction, is a transverse wave with speed $v_t = \sqrt{c_{44}/\rho}$.

[110] Direction

For propagation in the [110] direction $n_1 = n_2 = 1/\sqrt{2}$ and $n_3 = 0$ leading to

$$\begin{pmatrix} (c_{11}+c_{44})/2 - \rho v^2 & (c_{12}+c_{44})/2 & 0 \\ (c_{12}+c_{44})/2 & (c_{11}+c_{44})/2 - \rho v^2 & 0 \\ 0 & 0 & c_{44} - \rho v^2 \end{pmatrix} \begin{pmatrix} u_1^o \\ u_2^o \\ u_3^o \end{pmatrix} = 0. \quad (3.18)$$

Although the algebra is a bit more complicated in this case, setting the determinant of Equation 3.18 = 0 gives the results.

First eigenvalue,

$$v_l = \sqrt{\frac{c_{11}+c_{12}+2c_{44}}{2\rho}}. \quad (3.19)$$

Substituting this value for v into Equation 3.18 gives three equations,

$$\begin{aligned} -\left(\tfrac{c_{12}+c_{44}}{2}\right)u_1^o + \left(\tfrac{c_{12}+c_{44}}{2}\right)u_2^o &= 0 \\ +\left(\tfrac{c_{12}+c_{44}}{2}\right)u_1^o - \left(\tfrac{c_{12}+c_{44}}{2}\right)u_2^o &= 0 \\ -\tfrac{c_{11}+c_{12}}{2}u_3^o &= 0 \end{aligned} \quad (3.20)$$

which clearly shows that $u_1^o = u_2^0, u_3 = 0$. Thus, the displacement may be written as

$$\vec{u}_1 = \frac{u_1^o}{\sqrt{2}} \begin{pmatrix} 1 \\ 1 \\ 0 \end{pmatrix}. \quad (3.21)$$

The particle motion is in the [110] direction, the direction of propagation, so this is a longitudinal wave.

Second eigenvalue,

$$v_{t1} = \sqrt{\frac{c_{11}-c_{12}}{2\rho}}. \quad (3.22)$$

Substituting this value for v into 3.18 gives $u_1^o = -u_2^0, u_3 = 0$,

$$\vec{u}_2 = \frac{u_1^o}{\sqrt{2}} \begin{pmatrix} 1 \\ -1 \\ 0 \end{pmatrix}. \quad (3.23)$$

Third eigenvalue,

$$v_{t2} = \sqrt{\frac{c_{44}}{\rho}}. \tag{3.24}$$

Following the previous procedure gives $u_1^o = u_2^o = 0$ with no restriction on u_3^o so an acceptable solution is

$$\vec{u}_3 = u_3^o \begin{pmatrix} 0 \\ 0 \\ 1 \end{pmatrix}. \tag{3.25}$$

Taking dot products among $\vec{u}_1, \vec{u}_2$, and $\vec{u}_3$ shows that all three are mutually perpendicular, as they should be. In summary, for propagation in the [110] direction in a cubic crystal there is a pure longitudinal wave and two pure transverse waves polarized along $[1\bar{1}0]$ and [001] directions, respectively.

[111] Direction

For propagation in the [111] direction $n_1 = n_2 = n_3 = \frac{1}{\sqrt{3}}$. The results for this direction are, $v_l = \sqrt{\frac{c_{11}+2c_{12}+4c_{44}}{3\rho}}$ for longitudinal waves, and $v_{t1} = v_{t2} = \sqrt{\frac{c_{11}-c_{12}+c_{44}}{3\rho}}$ for transverse waves. The polarization of the transverse waves may be in any direction in a plane perpendicular to the [111] direction.

Pure mode waves are defined as those for which the displacement $\vec{u}$ (particle velocity) is parallel to the propagation direction (longitudinal) or perpendicular to the propagation direction (transverse).[4] There are three pure mode directions for a cubic crystal: [100]; [110]; and [111]. Normally, for other directions one wave has the particle velocity direction near the propagation direction (quasi-longitudinal) and two waves have the particle velocity directions almost perpendicular to the propagation direction (quasi-transverse). Such waves can be explored using the methods just described, although it would be fairly tedious to do by hand. A computer program is recommended.

3.1.2 Hexagonal Symmetry Solutions

The Christoffel determinant in this case becomes [33, 1],

$$\begin{vmatrix} n_1^2 c_{11} + n_2^2 \frac{c_{11}-c_{12}}{2} + n_3^2 c_{44} - \rho v^2 & n_1 n_2 \frac{c_{11}+c_{12}}{2} & n_1 n_3 (c_{13}+c_{44}) \\ n_1 n_2 \frac{c_{11}+c_{12}}{2} & n_1^2 \frac{c_{11}-c_{12}}{2} + n_2^2 c_{11} + n_3^2 c_{44} - \rho v^2 & n_2 n_3 (c_{13}+c_{44}) \\ n_1 n_3 (c_{13}+c_{44}) & n_2 n_3 (c_{13}+c_{44}) & n_3^2 c_{33} + (n_1^2+n_2^2) c_{44} - \rho v^2 \end{vmatrix} = 0. \tag{3.26}$$

[4] Another definition of pure mode is also in use [33].

[001] Direction

For this direction $n_1 = n_2 = 0, n_3 = 1$ and the Christoffel determinant becomes

$$\begin{vmatrix} c_{44} - \rho v^2 & 0 & 0 \\ 0 & c_{44} - \rho v^2 & 0 \\ 0 & 0 & c_{33} - \rho v^2 \end{vmatrix} = 0, \tag{3.27}$$

giving immediately the results: $v_l = \sqrt{c_{33}/\rho}$ for longitudinal waves along the [001] direction; and two solutions $v_t = \sqrt{c_{44}/\rho}$ for transverse waves polarized perpendicular to the [001] axis.

[100] Direction

In this case $n_1 = 1, n_2 = n_3 = 0$ and the Christoffel determinant again takes a simple form[5]

$$\begin{vmatrix} c_{11} - \rho v^2 & 0 & 0 \\ 0 & \frac{c_{11}-c_{12}}{2} - \rho v^2 & 0 \\ 0 & 0 & c_{44} - \rho v^2 \end{vmatrix} = 0, \tag{3.28}$$

resulting in $v_l = \sqrt{c_{11}/\rho}$ for longitudinal waves along the [100] direction; $v_{t1} = \sqrt{(c_{11} - c_{12})/2\rho}$ for transverse waves polarized along the [010] direction; and, $v_{t2} = \sqrt{c_{44}/\rho}$ for transverse waves polarized along the [001] direction.

*[**n**$_1$**n**$_2$0] Direction*

Although the algebra is a bit tedious, it is instructive to investigate the $[n_1 n_2 0]$ propagation direction. Taking into account $n_1^2 + n_2^2 = 1$, the three solutions for the velocities turn out to be independent of n_1 and n_2.

First eigenvalue There is a wave with $v_l = \sqrt{c_{11}/\rho}$ propagating in the $[n_1 n_2 0]$ direction. The displacement is characterized by $n_2 u_1^o = n_1 u_2^o$. The cross product between the wave vector and the displacement vector is zero,

$$\vec{K} \times \vec{u}^o = K(n_2 u_1^o - n_1 u_2^o) = 0 \tag{3.29}$$

so the displacement is parallel to the propagation direction; therefore the wave is longitudinal.

Second eigenvalue There is a wave with $v_{t1} = \sqrt{(c_{11} - c_{12})/2\rho}$. The displacement is characterized by $n_1 u_1^o = -n_2 u_2^o$. The dot product of the wave vector with the displacement vector is zero,

$$\vec{K} \cdot \vec{u}^o = K(n_1 u_1^o + n_2 u_2^o) = 0. \tag{3.30}$$

This is a transverse wave polarized in the plane perpendicular to the [001] direction.

[5] It is worth noting that the brackets, [ ], refer to an orthogonal coordinate system, not necessarily the crystalline axes. The *crystalline*, *a*, *b*, and *c* axes in the hexagonal system are not mutually orthogonal.

Third eigenvalue The third solution has $v_{t2} = \sqrt{c_{44}/\rho}$, a transverse wave polarized in the [001] direction.

The example discussed in this section shows that the properties of the elastic waves in a hexagonal crystal are *independent of propagation direction for propagation in a plane perpendicular to the* [001] *direction, the basal plane.* This result is consistent with the discussion of hexagonal crystals in Section 2.3.4.

$$\left[n_1 = \tfrac{1}{\sqrt{2}}, n_2 = 0, n_3 = \tfrac{1}{\sqrt{2}}\right] \textit{Direction}$$

A common practice is to determine the elastic constants of a crystal by measuring ultrasonic wave velocities along well-chosen directions. The preceding discussion has shown that four of the five hexagonal elastic constants, c_{11}, c_{12}, c_{33}, and c_{44} may be determined by measurements along the [001] and [100] directions. One of the constants, c_{13}, remains undetermined. This constant cannot be determined by measurements along such simple crystalline directions. A lower symmetry direction is needed. Any propagation direction intermediate between the [001] axis and the basal plane will involve c_{13} [34], and ($n_1 = n_3 = \frac{1}{\sqrt{2}}, n_2 = 0$) (at 45° to the x_1 and x_3 axes) is convenient. It is straight forward, if tedious, to solve for the eigenvalues and eigenvectors for this case.

First eigenvalue, $v_{1t} = \sqrt{\frac{c_{11}-c_{12}+2c_{44}}{4\rho}}$. This wave is polarized in the [010] direction and is a pure transverse wave.

Second and third eigenvalues,[6]

$$v_{2,3} = \sqrt{\frac{\frac{c_{11}+c_{33}+2c_{44}}{2} \pm \sqrt{\left(\frac{c_{11}-c_{33}}{2}\right)^2 + (c_{13}+c_{44})^2}}{2\rho}}, \tag{3.31}$$

with

$$\frac{u_1^o}{u_3^o} = \frac{1}{2}\left(\frac{c_{11}-c_{33}}{c_{13}+c_{44}}\right) \pm \sqrt{1 + \left(\frac{c_{11}-c_{33}}{2(c_{13}+c_{44})}\right)^2}. \tag{3.32}$$

If $c_{11} = c_{33}$, then $u_1^o = \pm u_3^o$ where the plus sign corresponds to a pure longitudinal mode with particle motion along the propagation direction, and the minus sign corresponds to a pure transverse mode polarized in the $[10\bar{1}]$ direction. However, in general $c_{11} \neq c_{33}$. As an example, using the elastic constants for the hexagonal crystal zinc [27] shows that the direction of the particle motion for the wave with the highest velocity of Equation 3.31 (the plus sign option in Equations 3.31 and 3.32) is approximately 34° above the basal plane, not 45°. Thus, this is a quasi-longitudinal wave. The other wave, the minus sign option, has the

[6] The reader should beware that several published results for this case have some apparent typographical errors. It is hoped that such is not the situation in the present work.

particle motion orthogonal to the quasi-longitudinal wave, and is thus a quasi-transverse wave.

3.1.3 Other Symmetries

In general, nonpure modes are required to determine at least some of the elastic constants as the symmetry is lowered. The same procedures used for the hexagonal case apply to the other cases, although the details will usually be more complicated.

Tetragonal

Tetragonal actually comprises two cases depending on point symmetry (Table 2.4): one with six independent elastic constants and a second with seven independent elastic constants. For the six elastic constant case, five of the constants are obtainable from pure mode propagation and the sixth requires a quasi-longitudinal or quasi-transverse mode [35]. The seven elastic constant case is considerably more difficult. Only two of the constants are obtained directly from pure modes. Several quasi modes are required to complete the set [36].

Orthorhombic

Six of the nine independent elastic constants are obtained from pure mode propagation along simple directions. The other three require nonpure modes along three different directions [37].

3.1.4 Phase Velocity and Group Velocity

Plane waves were introduced *via* Equation 3.4 using a complex exponential representation for the wave displacement $\vec{u}$. The complex exponential function is very convenient for mathematical manipulations, but if a physical value of the wave amplitude is needed, the real part of $\vec{u}$ is used. As a special case for the following discussion, we will assume that the wave travels in the $\hat{x}_1$ direction, ($\vec{K} = K\hat{x}_1$) and we will look only at the $\hat{x}_2$ component of the displacement. Taking the real part leads to

$$u_2(x_1,t) = u_2^o \cos(Kx_1 - \omega t). \tag{3.33}$$

The value of $u_2(x_1,t)$ will repeat as the argument $(Kx_1 - \omega t)$ changes by 2π. This could occur by a change in time of $T = 2\pi/\omega$, where T is the period of the wave, by a change in space of a distance $\lambda = 2\pi/K$ where λ is the wavelength, or by a combination of the two. Starting from some initial reference values of x_1 and t, the fraction of the wave cycle (2π) that the wave has advanced is called the *phase* of the wave. Now imagine moving along the direction of propagation of the wave at a velocity such that the local observed displacement of the wave is constant. Such a velocity is called the *phase velocity* and results when $Kx_1 - \omega t = CONSTANT$. Taking the derivative with respect to time gives the phase velocity, $dx_1/dt = \omega/K = v$. This relation is usefully written,

$$\omega = vK. \tag{3.34}$$

The velocities calculated in earlier parts of Section 3.1 are phase velocities.

Phase velocity describes the speed of single frequency plane waves as given by Equation 3.4. These waves are an idealization, and in fact, such waves transmit no information. However, these waves can be used to build wave packets, and these packets find many uses. Wave packets travel at the *group velocity* as will now be shown [33, 38]. A wave packet may be built up from plane waves

$$U(x_1, t) = \frac{1}{\sqrt{2\pi}} \int_{-\infty}^{\infty} A(K) \exp\left[i(Kx_1 - \omega t)\right] dK \tag{3.35}$$

where $A(K)$ gives the amplitude of the various plane waves in the packet and ω is assumed to be a function of K. Assuming that $A(K)$ is rather sharply peaked about the value K_o, an expansion is possible,

$$\omega(K) \simeq \omega_o + \left(\frac{d\omega}{dK}\right)_o (K - K_o), \tag{3.36}$$

where $\omega_o = \omega(K_o)$. In anticipation of the final result, define $v_g = \left(\frac{d\omega}{dK}\right)_o$. Equation 3.35 becomes

$$U(x_1, t) = \frac{\exp\left[i(K_o v_g - \omega_o)t\right]}{\sqrt{2\pi}} \int_{-\infty}^{\infty} A(K) \exp\left[iK(x_1 - v_g t)\right] dK. \tag{3.37}$$

Next is a bit of a tricky move. From Equation 3.35,

$$U(x_1, 0) = \frac{1}{\sqrt{2\pi}} \int_{-\infty}^{\infty} A(K) \exp\left[i(Kx_1)\right] dK. \tag{3.38}$$

Next, define $x_1' = x_1 - v_g t$. The integral in Equation 3.37 is just $U(x_1', 0)$. Combining these equations results in

$$U(x_1, t) = U(x_1', 0) \exp\left[i(K_o v_g - \omega_o)t\right]. \tag{3.39}$$

Except for a phase factor, the wave packet at time t is just the packet at time 0, but shifted a distance $v_g t$ along the x_1 axes. Thus, within the approximation of Equation 3.36, the packet moves with the unchanged shape at a velocity v_g, the group velocity. The group velocity is also equal to the energy transport velocity [33].

For the one-dimensional problem just discussed, the group velocity of a wave packet is given by $v_g = d\omega/dK$. More generally $\vec{v}_g$ is the gradient of ω in K space [39],

$$\vec{v}_g = \nabla_K \omega(\vec{K}). \tag{3.40}$$

3.2 Lattice Dynamics

3.2.1 Introduction

Although many of the vibrational properties of matter can be explained by treating matter as a continuum, much is missed. It is no surprise that the atomic nature of matter is necessary to explain many short-wavelength vibrational phenomena unaccounted for by the continuum model. The present section deals with contributions to the vibrational properties of materials arising from their crystalline nature.[7] Some knowledge of crystallography would be helpful for this section as it would be for other parts of the book; however an attempt is made to present the key points without resorting to extensive crystallographic details. Good discussions of crystallography are available in several standard books on solid state physics [39, 40, 41]. A rather detailed discussion is presented in the book by Burns [42].

Simple 1D models will be treated first, before taking up the 3D case. Some of the approximations intrinsic to the approach used are not discussed in detail in the present treatment. Fuller discussions are found elsewhere [43, 44].

3.2.2 Simple 1D Models

Monatomic Case

Figure 3.1 shows a simple model for a one-dimensional lattice that is relatively easy to solve [39, 40, 41, 45], and illustrates several important points which are qualitatively similar to those found in more realistic models. Atoms of mass M are shown in their equilibrium positions a distance a apart. They are labeled by $s-2, s-1$, *etc.* and nearest neighbors are connected by springs of force constant C. The use of springs in the illustration implies that the forces acting on the atoms are linearly related to the displacements from equilibrium (Hooke's Law behavior), which in turn implies that the interaction energy depends quadratically on these displacements. In practice the interaction energy is usually expanded in a Taylor series in the displacements. The linear terms vanish for displacements relative to

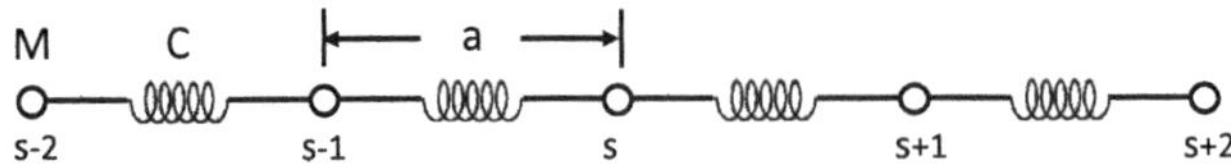

Figure 3.1 Simple model of a 1-D lattice with atoms connected by springs. The mass of each atom is M and the spring constant is C. The atoms are spaced a distance a apart and labeled by $s-2, s-1$, etc.

[7] Quasicrystalline materials will not be discussed in this Section.

the equilibrium positions and only the quadratic terms are usually used as a good approximation for small-amplitude vibrations.

The equations of motion for these atoms will now be developed. The departures from equilibrium are labeled u_{s-1}, u_s, u_{s+1}, *etc.* The force acting on the sth atom is

$$F_s = C(u_{s+1} - u_s) - C(u_s - u_{s-1}). \quad (3.41)$$

If $u_{s+1} > u_s$, the first term represents a force to the right due to the stretch of the spring between u_{s+1} and u_s, while the second term is a force to the left if $u_s > u_{s-1}$. Reversing the inequalities corresponds to compressions of the springs and changes the signs of the forces. The equation of motion is

$$M\frac{d^2u_s}{dt^2} = C(u_{s+1} + u_{s-1} - 2u_s). \quad (3.42)$$

Substituting the trial solution,

$$u_s = A\exp\left[i(Kas - \omega t)\right], \quad (3.43)$$

into Equation 3.42 and noting that $u_{s+1} = \exp(iKa)u_s$ with a corresponding relation for u_{s-1}, results in

$$-M\omega^2 = C\Big[\exp(iKa) + \exp(-iKa) - 2\Big], \quad (3.44)$$

where a common factor of u_s has been divided out. The Euler formula for complex exponentials gives the result,

$$\omega^2 = \frac{2C}{M}(1 - \cos(Ka)). \quad (3.45)$$

This equation is more commonly written

$$\omega = 2\sqrt{\frac{C}{M}}|\sin Ka/2|. \quad (3.46)$$

Only positive values are of interest for the frequency. Equation 3.46, relating ω to K, is known as a dispersion relation.

Several conclusions of general significance can be drawn from these results.

i. The dispersion relation, Equation 3.46 and Figure 3.2, for the model of atoms connected by springs is very different than that found for long-wavelength acoustic waves, Equation 3.34. This difference is due to the discrete, periodic nature of the crystalline lattice. However, for $Ka = 2\pi a/\lambda << 1$, the regime of validity of the continuum approach, the two results agree, ω depends linearly on K.

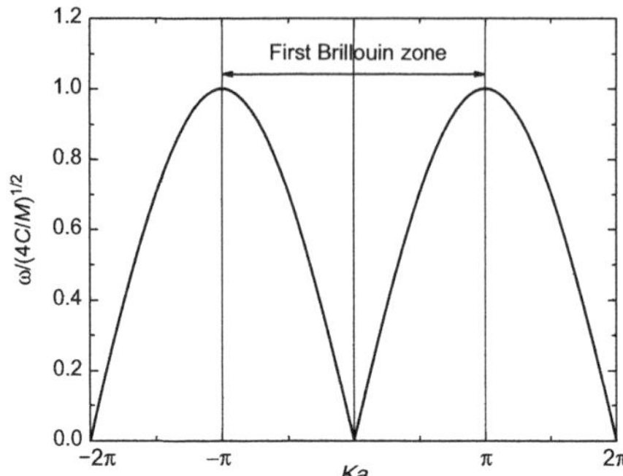

Figure 3.2 Plot of ω *vs.* Ka, the dispersion curve for the simple 1D lattice of atoms coupled by springs.

ii. The group velocity, $v_g = d\omega/dt$, is very different for the two cases. For $Ka << 1$ the results agree, v_g is independent of K. However, as Figure 3.2 shows, v_g decreases for higher values of Ka for the 1D lattice, and goes to zero at $Ka = \pi$. This effect can be understood from the following consideration. Substituting $K = 2\pi/\lambda$ shows that $\lambda = 2a$ at this point. This condition is just the requirement for constructive interference for waves reflected from two adjacent atoms. Waves traveling to the right satisfy the Bragg condition [39] and are converted to waves traveling to the left. However, waves traveling to the left face the same condition. The result is a standing wave, not a traveling wave.

iii. Equation 3.2 repeats indefinitely along the K axis. However, all the unique solutions for the particle displacement u_s are found in a Ka range of 2π. Usually, this range is chosen as $-\pi \le Ka \le +\pi$. This range is called the first Brillouin zone, as indicated in Figure 3.2. Suppose there is a solution somewhere outside the first Brillouin zone in Figure 3.2. Call this value of the wave vector K'. There is some value of an integer n for which $K' = K + 2\pi n/a$ where K is in the first Brillouin zone.

Then

$$\begin{aligned} u_s(K') &= A\exp[i(K'as - \omega t)] = A\exp[i((K + 2\pi n/a)as - \omega t)] \\ &= A\exp[i(Kas + 2\pi ns - \omega t)] = u_s(K). \end{aligned} \tag{3.47}$$

The last step is only true if s is an integer, but integer s *defines* the atomic positions. Thus all the unique solutions for the displacement of the atoms can be found within the first Brillouin zone. Graphical illustrations of this situation are found in several places [39, 42, 45]. A similar situation holds in 3D, all unique solutions for the atomic displacements may be found in the first Brillouin zone, which in 3D is a cell in K space.

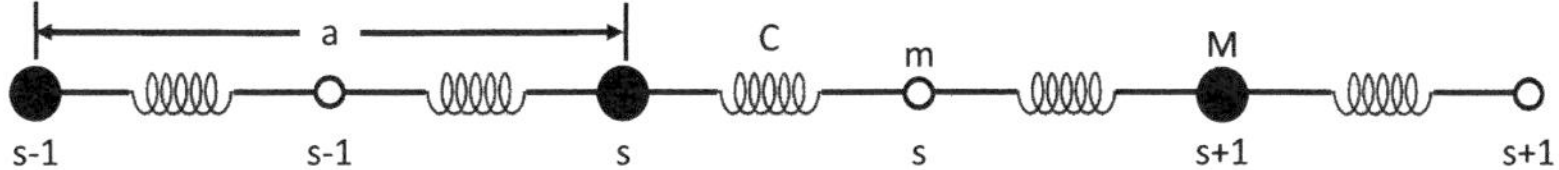

Figure 3.3 Simple model of a 1D lattice with two atoms per unit cell. The mass of the atoms are M and m, and the spring constant is C for all springs. Like atoms are spaced a distance a apart and labeled by $s-1, s$, etc.

Diatomic Case

New features emerge when the case of two atoms per unit cell is examined [43]. The situation is illustrated in Figure 3.3.

The displacements from equilibrium will be labeled u_s for M and v_s for m. Similar arguments as those leading to 3.42, give the equations of motion,

$$\begin{aligned} M\frac{d^2u_s}{dt^2} &= C(v_s + v_{s-1} - 2u_s) \\ m\frac{d^2v_s}{dt^2} &= C(u_{s+1} + u_s - 2v_s). \end{aligned} \tag{3.48}$$

Substituting the trial solutions

$$\begin{aligned} u_s &= A\exp[i(Kas - \omega t)] \\ v_s &= B\exp[i(Kas - \omega t)] \end{aligned} \tag{3.49}$$

into Equation 3.48, noticing that $v_{s-1} = \exp[-iKa]v_s$ with a related expression for u_{s+1}, and dividing out common exponential factors finally yields two linear equations for A and B, which are conveniently written in matrix form,

$$\begin{pmatrix} 2C - M\omega^2 & -C(1+\exp(-iKa)) \\ -C(1+\exp(iKa)) & 2C - m\omega^2 \end{pmatrix} \begin{pmatrix} A \\ B \end{pmatrix} = 0. \tag{3.50}$$

Setting the determinant equal to zero gives the relation between ω and K, the dispersion relation,

$$\omega = \sqrt{C\left(\frac{1}{M}+\frac{1}{m}\right) \pm C\sqrt{\left(\frac{1}{M}+\frac{1}{m}\right)^2 - 4\frac{\sin^2(Ka/2)}{Mm}}}. \tag{3.51}$$

An essential feature of the diatomic case is that there are two branches of Equation 3.51, corresponding to the $\pm$ sign, as shown in Figure 3.4. The two branches are called “acoustic” and “optic” as will be discussed below. As was the case for Figure 3.2, all unique solutions for the particle motion are represented in the first Brillouin zone. Equation 3.51 is a bit complicated, so it is instructive to look at special cases.

First, consider the center region of the Brillouin zone. For $Ka << 1$, $\sin(Ka/2) \approx Ka/2$. Expanding Equation 3.51 to get the small Ka results and substituting these into Equation 3.50 to get the eigenvectors shows that

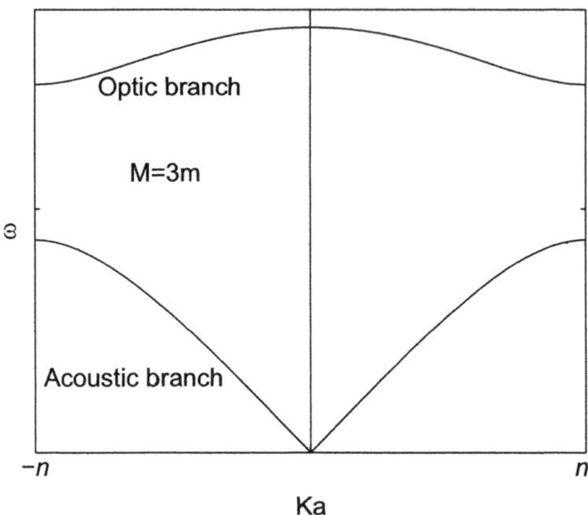

Figure 3.4 Dispersion curve for the diatomic 1D chain with $M = 3m$. Only the first Brillouin zone is shown. The labels "Optic branch" and "Acoustic branch" are discussed in the text.

1. **Optic branch**

$$\omega \approx \sqrt{2C\left(\frac{1}{M}+\frac{1}{m}\right)} \qquad \frac{A}{B} \approx -\frac{m}{M} \tag{3.52}$$

2. **Acoustic branch**

$$\omega \approx \sqrt{\frac{C}{2(M+m)}}Ka \qquad \frac{A}{B} \approx 1. \tag{3.53}$$

The acoustic branch is so named because, in the long-wavelength limit (small K), ω depends linearly on K, as was found earlier using the Christoffel equations for long-wavelength acoustic waves. The vibrational amplitudes of adjacent atoms of masses M and m are approximately equal. The situation is quite different for the optic branch. The frequency does not go to zero in the long-wavelength limit. Adjacent atoms vibrate in opposite directions, in fact they vibrate about their center of mass. If the atoms have opposite charges, the result is an oscillating electric dipole moment which couples to electromagnetic waves, *e.g.*, light or infrared radiation, hence the name optic branch.

Next, the Brillouin zone boundary is considered. At $Ka = \pm\pi$, $\sin^2(Ka/2) = 1$ and Equation 3.51 reduces to

$$\omega = \sqrt{C\left(\frac{1}{M}+\frac{1}{m}\right) \pm \left(\frac{1}{M}-\frac{1}{m}\right)}. \tag{3.54}$$

Substituting each solution in turn into Equation 3.50 gives the eigenvectors. Then,

1. **Optic branch**

$$\omega \approx \sqrt{2C/m} \qquad A = 0 \quad B \neq 0 \tag{3.55}$$

2. **Acoustic branch**

$$\omega \approx \sqrt{2C/M} \qquad A \neq 0 \quad B = 0. \tag{3.56}$$

Notice that at the zone boundary only the atoms of mass m move for the optic mode, which is consistent with the dependence of the frequency only on mass m. Similar comments apply to the acoustic mode. Illustrations of the atomic movements are to be found in various books on solid state physics [39, 40, 41, 42].

3.2.3 Lattice Dynamics in Three Dimensions

Introduction

An introduction to lattice dynamics in three dimensions will be given in the present section. The basic ideas are similar to the 1D cases discussed in Section 3.2.2, but the notation tends to get a bit cumbersome. The reader should not be discouraged; the basic ideas are simple [41, 45, 46], and every effort will be made to clearly define the terms used in the discussion. The potential energy is assumed to be a function of the instantaneous positions of all the atoms in the crystal. This energy will be expanded in a Taylor series in terms of the displacements of the atoms from their equilibrium positions. Only terms through the second order will be kept, which is the harmonic approximation. It is necessary to define the atomic displacement coordinates clearly.

First, a reminder of a few concepts regarding crystal lattices. The description of a crystal lattice usually starts with the concept of a Bravais lattice, which is an infinite array of mathematical points filling space in a periodic manner. There are 14 distinct Bravais latttice types classified according to their symmetry properties. These 14 types are grouped into seven systems. A basis is added to each Bravais lattice point to form a crystal lattice. Each basis is an arrangement of atoms that is identical in every way, including orientation, at each Bravais latttice point. In some cases the basis consists of a single atom in which case the Bravais lattice and the crystal are the same structure. More generally, and for many interesting cases, the basis consists of more than one atom.

Choosing as origin of a coordinate system a particular Bravais lattice point, the lattice vector $\vec{R}_l$ illustrated in Figure 3.5 indicates the position of the lth Bravais lattice point. It is assumed that there is a basis associated with the lattice point; $\vec{d}_n$ defines the location of the nth atom in the basis. The displacement from equilibrium of the atom located at $\vec{R}_l + \vec{d}_n$ is indicated by $\vec{u}(n,l)$. The number n typically ranges

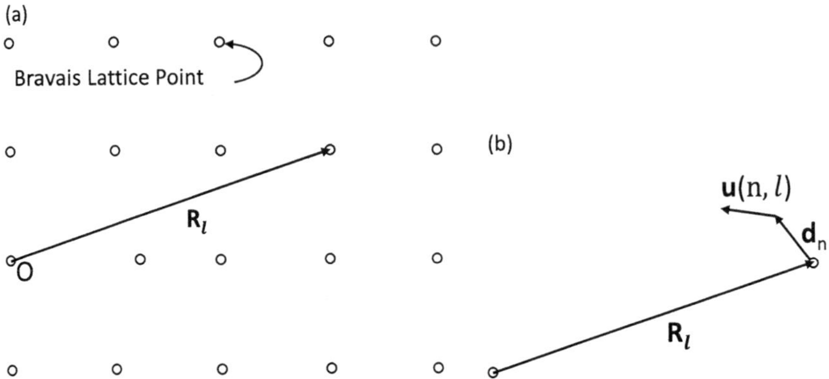

Figure 3.5 (a) An illustration of the points in a Bravais lattice. The vector $\vec{R}_l$ shows the lth Bravais lattice point relative to the origin indicated, where l ranges over all the Bravais lattice points in the crystal. (b) Although not indicated in Part (a), it is assumed that there is a basis associated with each Bravais lattice point. The vector $\vec{d}_n$ gives the location of the nth atom in basis. Finally $\vec{u}(n, l)$ gives the displacement of the nth atom in the lth unit cell.

from one to a small number. In cases where there is only one atom associated with a Bravais lattice point, the concept of a basis is superfluous and is ignored.

Taylor Series for the Potential Energy

The potential energy V is expanded in terms of the displacements of the atoms from their equilibrium positions.

$$
\begin{aligned}
V \approx V_o &+ \sum_{nl\alpha} \left(\frac{\partial V}{\partial u_\alpha(n,l)} \right)_o u_\alpha(n,l) \\
&+ \frac{1}{2} \sum_{nl\alpha} \sum_{n'l'\beta} \left(\frac{\partial^2 V}{\partial u_\alpha(n,l) \partial u_\beta(n',l')} \right)_o u_\alpha(n,l) u_\beta(n',l') + \cdots
\end{aligned}
\tag{3.57}
$$

where α and β each range from 1 to 3, the three x_i coordinate directions. The first term in Equation 3.57 is the cohesive energy, and is not of interest for the present problem. The second term, related to the forces on the atoms at their equilibrium positions is zero. Equation 3.57 is written as

$$
V \approx V_o + \frac{1}{2} \sum_{nl\alpha} \sum_{n'l'\beta} \Phi_{\alpha\beta}^{nl,n'l'} u_\alpha(n,l) u_\beta(n',l'), \tag{3.58}
$$

with

$$
\Phi_{\alpha\beta}^{(nl,n'l')} = \left(\frac{\partial^2 V}{\partial u_\alpha(n,l) \partial u_\beta(n',l')} \right)_o . \tag{3.59}
$$

Although the notation looks complex, it may help to remember that $(nl, n'l')$ simply refers to a pair of atoms,with one location denoted by (nl) and the other by $(n'l')$. Specific pairs will be considered later. In the examples to be considered the pairs are usually nearest-neighbors or second nearest-neighbors.

The Force Constant Matrix and the Equations of Motion

The next step is to find the forces on the atoms due to their relative displacements and then to get the equation of motion. Consider a particular cartesian component of the motion, β', for a particular atom (n'', l''). Then,

$$F_{\beta'}(n'', l'') = -\frac{\partial V}{\partial u_{\beta'}(n'', l'')}. \tag{3.60}$$

In evaluating the derivative it is important to remember that the set (n'', l'', β') refers to a *particular* atom and *particular* coordinate while each sum in Equation 3.58 ranges over *all* atoms and coordinates including (n'', l'', β'). The result is

$$F_{\beta'}(n'', l'') = -\sum_{n'l'\beta} \Phi_{\beta'\beta}^{(n''l'',n'l')} u_\beta(n', l'). \tag{3.61}$$

However, having introduced the (n'', l'', β') notation, it is no longer needed. Making the change $l'' \to l$, *etc.*

$$F_\alpha(n, l) = -\sum_{n'l'\beta} \Phi_{\alpha\beta}^{(nl,n'l')} u_\beta(n', l'). \tag{3.62}$$

The interpretation of the force constant matrix, Φ, is now easy. Each term in the sum is the α component of the force on the atom at (n, l) due to the β component of the displacement of the atom at (n', l'). The quantity Φ relates the force on one atom to the displacement of another, and is called the force constant, often expressed as a matrix.

The equation of motion for an atom of mass M_n located at (n, l) is

$$M_n \ddot{u}_\alpha(n, l) = -\sum_{n'l'\beta} \Phi_{\alpha\beta}^{(nl,n'l')} u_\beta(n', l'). \tag{3.63}$$

Suppose there are N_b atoms in the basis. Then $\alpha = 1, 2, 3$ while n ranges from 1 to N_b, thus Equation 3.63 results in $3N_b$ total equations.

Plane Wave Solutions and the Dynamical Matrix

Solutions of the form

$$u_\alpha(n, l, t) = A_\alpha(n) \exp[i(\vec{K} \cdot \vec{R}_l - \omega t)] \tag{3.64}$$

will be substituted into Equation 3.63. Making the substitution, dividing by the common factor $\exp(i\omega t)$ and multiplying both sides by $\exp(-i\vec{K}\cdot\vec{R}_l)$ gives

$$M_n\omega^2 A_\alpha(n) = \sum_{n'l'\beta} \Phi_{\alpha\beta}^{(nl,n'l')} \exp\left(i\vec{K}\cdot(\vec{R}_{l'}-\vec{R}_l)\right) A_\beta(n'). \tag{3.65}$$

Notice that the result only depends on the *separation* of lattice points $(R_{l'} - R_l)$, as one would expect, not their absolute position [45]. That fact will be used below to set $l = 0$, a convenient reference lattice point.

It might seem that R_l and $R_{l'}$ in Equation 3.65 should instead be $\vec{R}_l + \vec{d}_n$ and $\vec{R}_{l'}+\vec{d}_{n'}$ respectively, the positions of atoms (n, l) and (n', l'). The difference between these two choices simply gets absorbed into the phase of $A_\beta(n')$ and has no effect on the calculated values for ω^2 [45, 47].

Equation 3.65 can be written in a somewhat more standard form by defining $A_\alpha(n) = \frac{1}{\sqrt{M_n}}e_\alpha(n)$. Equation 3.65 becomes

$$\omega^2 e_\alpha(n) = \sum_{n'\beta} D_{\alpha\beta}(n,n')e_\beta(n'), \tag{3.66}$$

where

$$D_{\alpha\beta}(n,n') = \frac{1}{\sqrt{M_n M_n'}} \sum_{l'} \Phi_{\alpha\beta}^{(n0,n'l')} \exp\left(i\vec{K}\cdot\left(\vec{R}_{l'}-\vec{R}_0\right)\right) \tag{3.67}$$

is called the dynamical matrix. In Equation 3.67, l has been chosen as 0. Equation 3.66 is a set of $3N_b$ linear equations, which may be solved by standard methods for the eigenvalues, ω^2, and eigenvectors, $e_\alpha(n)$.

It might be helpful to rewrite Equation 3.66 as

$$\sum_{n'\beta} \left(D_{\alpha\beta}(n,n') - \delta_{\alpha\beta}\delta_{nn'}\omega^2\right) e_\beta(n') = 0. \tag{3.68}$$

The nature of the problem has resulted in a (perhaps) confusing set of indices. It may be helpful to choose a simplified case for illustration. Suppose there are two atoms in the basis. Then n takes on the values 1 or 2 while α ranges from 1 to 3. A term such as $D_{13}(1,2)$ is the xz component[8] of the dynamical matrix connecting atom 1 to atom 2 in the basis. The number of neighboring Bravais lattice points to include is controlled by the sum over l' in $D_{\alpha\beta}(n,n')$, Equation 3.67. Equation 3.68 may be written in matrix form. There is no required order in which to list the elements in the matrix, but the following choice seems reasonable [41, 45]: first list the $\alpha = 1,2,3$ components of displacement for atom 1 of the basis, then do the same thing for atom 2, *etc.*

[8] It seems better to temporarily switch from x_1, x_2, x_3 notation to x, y, z notation.

$$\begin{pmatrix} D_{xx}(11)-\omega^2 & D_{xy}(11) & D_{xz}(11) & D_{xx}(12) & D_{xy}(12) & D_{xz}(12) \\ D_{yx}(11) & D_{yy}(11)-\omega^2 & D_{yz}(11) & D_{yx}(12) & D_{yy}(12) & D_{yz}(12) \\ D_{zx}(11) & D_{zy}(11) & D_{zz}(11)-\omega^2 & D_{zx}(12) & D_{zy}(12) & D_{zz}(12) \\ D_{xx}(21) & D_{xy}(21) & D_{xz}(21) & D_{xx}(22)-\omega^2 & D_{xy}(22) & D_{xz}(22) \\ D_{yx}(21) & D_{yy}(21) & D_{yz}(21) & D_{yx}(22) & D_{yy}(22)-\omega^2 & D_{yz}(22) \\ D_{zx}(21) & D_{zy}(21) & D_{zz}(21) & D_{zx}(22) & D_{zy}(22) & D_{zz}(22)-\omega^2 \end{pmatrix} * \begin{pmatrix} e_x(1) \\ e_y(1) \\ e_z(1) \\ e_x(2) \\ e_y(2) \\ e_z(2) \end{pmatrix} = 0. \tag{3.69}$$

Properties of the Force Constant Matrix

Several key features of the force constant matrix will now be summarized

i. From Equation 3.62 it is clear that $-\Phi_{\alpha\beta}^{(nl,n'l')}u_\beta(n',l')$ is the α component of the *force* on an atom at (n,l) due to the β component of the *displacement* of an atom at (n',l'). In principle, $\Phi_{\alpha\beta}^{(nl,n'l')}$ can be calculated from basic physics, or limits can be placed on its form for central forces; however, for the present work, the $\Phi_{\alpha\beta}^{(nl,n'l')}$ will simply be taken as parameters of the lattice dynamics theory.
ii. From Equation 3.59,

$$\Phi_{\alpha\beta}^{(nl,n'l')} = \Phi_{\beta\alpha}^{(n'l',nl)} \tag{3.70}$$

because the order of differentiation makes no difference.
iii. If all atoms are displaced equally (the specimen is simply shifted a constant distance) there is no force on the atoms. Suppose each u_β in Equation 3.62 is replaced by a constant value, d_β. With $F_\alpha(n,l) = 0$, the arbitrary constant can be factored out with the result

$$\sum_{n'l'} \Phi_{\alpha\beta}^{(nl,n'l')} = 0, \tag{3.71}$$

a very important sum rule.
iv. The force constant matrix has certain symmetry properties, which may be used to simplify calculations. There is a difference between the case of one atom per unit cell located on a Bravais lattice point, and the case of multiple atoms per unit cell occupying a basis.

First, the case of one atom per unit cell is considered. Then $\Phi_{\alpha\beta}(\vec{R}_l - \vec{R}_{l'}) = \Phi_{\alpha\beta}(\vec{R}_{l'} - \vec{R}_l)$, which may be expressed as

$$\Phi_{\alpha\beta}^{(l,l')} = \Phi_{\alpha\beta}^{(l',l)}. \tag{3.72}$$

This follows from the fact that each Bravais lattice point is a point of inversion symmetry [40]. Combining Equation 3.70 and 3.72 gives

$$\Phi^{(l,l')}_{\alpha\beta} = \Phi^{(l,l')}_{\beta\alpha}. \tag{3.73}$$

The force constant matrix is symmetric in $\alpha\beta$.

For a crystal with multiple atoms per unit cell, the basis points are not normally at a point of inversion symmetry. Thus, the previous argument does not apply. However, *if* $\Phi^{(nl,n'l')}_{\alpha\beta}$ depends only on the distance between the atoms (n, l) and (n', l') *i.e.* central forces, then the $\alpha\beta$ symmetry applies [48, 49]

$$\Phi^{(nl,n'l')}_{\alpha\beta} = \Phi^{(nl,n'l')}_{\beta\alpha}. \tag{3.74}$$

Example: fcc System

The first step is to determine the force constant matrices. A face-centered cubic (fcc) lattice as illustrated in Figure 3.6 will be used to explore some features of lattice dynamics in 3-D. The fcc lattice is a Bravais lattice. The example being considered has only one atom per Bravais lattice point. For a particular pair of atoms the force-constants between them can be represented by a 3 by 3 matrix,

$$\Phi^{0,l'} = \begin{pmatrix} \phi_{xx} & \phi_{xy} & \phi_{xz} \\ \phi_{xy} & \phi_{yy} & \phi_{yz} \\ \phi_{xz} & \phi_{yz} & \phi_{zz} \end{pmatrix}. \tag{3.75}$$

By Equation 3.73 the matrix is symmetrical. Because there is only one atom per basis the n has been dropped. Also, the atom at (000) is labeled $l = 0$. The task now is to find 13 force constant matrices, those for the 12 nearest neighbors *and* that for the atom at (000).

First, the atom at (110) is considered. Symmetry arguments are used to simplify the general matrix of Equation 3.75. Relevant symmetry operations leave the fcc

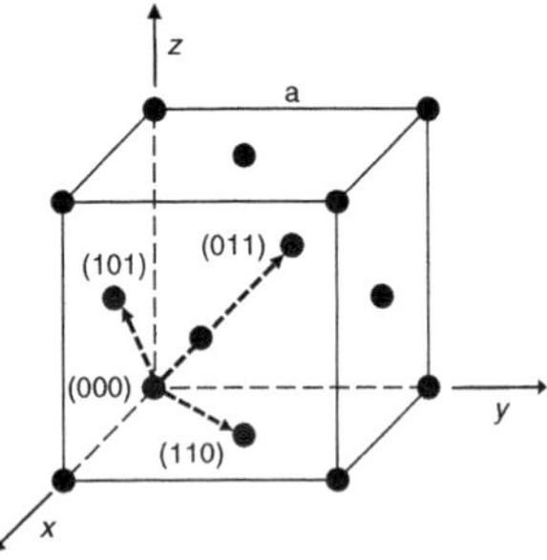

Figure 3.6 The conventional cell of a face-centered cubic lattice is illustrated, with atoms at each corner of the cube and in the centers of the faces. The force constant matrices between the atom located at (000) and its nearest neighbors will be found. In terms of $a/2$ the nearest neighbors are located at $\pm(110)$, $\pm(101)$, $\pm(011)$, $\pm(\bar{1}10)$, $\pm(\bar{1}01)$, and $\pm(0\bar{1}1)$. Only three of the 12 nearest neighbors are shown.

lattice invariant and, for the present case, leave the atoms at (000) and (110) in place. Equation 3.59 shows that the $\phi_{\alpha\beta}$ in Equation 3.75 involve second derivatives. Consider a reflection in the $x-y$ plane: $x \to x$, $y \to y$, $z \to -z$. Then $\frac{\partial}{x} \to \frac{\partial}{x}$, $\frac{\partial}{y} \to \frac{\partial}{y}, \frac{\partial}{z} \to -\frac{\partial}{z}$. Thus

$$
\begin{aligned}
\phi_{xz} = \frac{\partial^2 V}{\partial x \partial z} \to -\frac{\partial^2 V}{\partial x \partial z} &= -\phi_{xz}, \\
\phi_{xz} &= -\phi_{xz}, \\
\phi_{xz} &= 0.
\end{aligned}
\tag{3.76}
$$

The same type of argument shows that $\phi_{yz} = 0$.

Next, consider a reflection in a vertical plane passing through a line in the (110) direction. In this case $x \to y$ and $y \to x$ with the result that $\frac{\partial^2 V}{\partial x^2} \to \frac{\partial^2 V}{\partial y^2}$, or $\phi_{xx} = \phi_{yy}$. The six independent parameters of the matrix of Equation 3.75 have been reduced to three. It is convenient to write this matrix[9, 10] [50]

$$
\Phi^{0,110} = -\begin{pmatrix} \alpha & \gamma & 0 \\ \gamma & \alpha & 0 \\ 0 & 0 & \beta \end{pmatrix}. \tag{3.77}
$$

Because the Bravais lattice has inversion symmetry this result for the atom at (110) also applies for the atom at $-(110) = (\bar{1}\bar{1}0)$.

Other matrices may be found by applying the appropriate symmetry operations. For example, the matrix

$$
R_x(\theta) = \begin{pmatrix} 1 & 0 & 0 \\ 0 & \cos(\theta) & -\sin(\theta) \\ 0 & \sin(\theta) & \cos(\theta) \end{pmatrix}, \tag{3.78}
$$

will rotate a *vector* counterclockwise by an angle θ about the x axis. As an example, consider a vector from the origin to the (110) position, and consider $\theta = \pi/2$. Then the new vector is

$$
\begin{pmatrix} 1 & 0 & 0 \\ 0 & 0 & -1 \\ 0 & 1 & 0 \end{pmatrix} \begin{pmatrix} 1 \\ 1 \\ 0 \end{pmatrix} = \begin{pmatrix} 1 \\ 0 \\ 1 \end{pmatrix}, \tag{3.79}
$$

as is evident from Figure 3.6. To rotate the *matrix* $\Phi^{0,110}$ through the same angle the appropriate operaton is [9]

[9] The labels α and β for the force constants, in common usage, should not be confused with the $\alpha\beta$ subscripts for the Cartesian coordinates.

[10] As an example, by Equation 3.62 the xy term, γ, relates the x component of the force on the atom at (000) to the y component of the displacement of the atom at (110).

$$R(\pi/2) * \Phi^{0,110} * R^T(\pi/2) = -\begin{pmatrix} \alpha & 0 & \gamma \\ 0 & \beta & 0 \\ \gamma & 0 & \alpha \end{pmatrix}. \tag{3.80}$$

where R^T is the transpose of R. Because the fcc crystal is invariant under this rotation, the resulting force constant matrix is that for the pair of atoms at (000) and (101). Appropriate rotations about the x, y, or z axes yield the force constant matrices for all 12 nearest neighbors of the atom at (000) and are listed in Table 3.1 [46, 50]. Note that the forms of the matrices are determined by symmetry arguments. The last entry in Table 3.1 is found using Equation 3.71, the sum rule.

Having determined the form of the force constant matrices for the fcc case, the next step is to find the dynamical matrix using Equation 3.67 and Table 3.1. It is helpful to note that the atoms occur in pairs, indicated by the $\pm$ signs in Table 3.1. The exponential terms in Equation 3.67 can be simplified using the relation $e^{ix} + e^{-ix} = 2\cos(x)$. The n and n' in Equation 3.67 will be supressed because there

Table 3.1 *Force constant matrices* $\Phi_{\alpha\beta}^{(0,l')}$ *for the fcc lattice.*

Atom positions in terms of $a/2$	Force constant matrix
$\pm(110)$	$-\begin{pmatrix} \alpha & \gamma & 0 \\ \gamma & \alpha & 0 \\ 0 & 0 & \beta \end{pmatrix}$
$\pm(\bar{1}10)$	$-\begin{pmatrix} \alpha & -\gamma & 0 \\ -\gamma & \alpha & 0 \\ 0 & 0 & \beta \end{pmatrix}$
$\pm(101)$	$-\begin{pmatrix} \alpha & 0 & \gamma \\ 0 & \beta & 0 \\ \gamma & 0 & \alpha \end{pmatrix}$
$\pm(\bar{1}01)$	$-\begin{pmatrix} \alpha & 0 & -\gamma \\ 0 & \beta & 0 \\ -\gamma & 0 & \alpha \end{pmatrix}$
$\pm(011)$	$-\begin{pmatrix} \beta & 0 & 0 \\ 0 & \alpha & \gamma \\ 0 & \gamma & \alpha \end{pmatrix}$
$\pm(0\bar{1}1)$	$-\begin{pmatrix} \beta & 0 & 0 \\ 0 & \alpha & -\gamma \\ 0 & -\gamma & \alpha \end{pmatrix}$
(000)	$\begin{pmatrix} 8\alpha + 4\beta & 0 & 0 \\ 0 & 8\alpha + 4\beta & 0 \\ 0 & 0 & 8\alpha + 4\beta \end{pmatrix}$

is only one atom per basis in the present example. The l' sum is over the six nearest neighbors plus the atom at (000). The first term is

$$\begin{aligned} D_{xx}(\vec{K}) = \frac{1}{M}\Big[&- 2\alpha \cos\left(\vec{K}\cdot\frac{a}{2}(\hat{x}+\hat{y})\right) - 2\alpha \cos\left(\vec{K}\cdot\frac{a}{2}(-\hat{x}+\hat{y})\right) \\ &- 2\alpha \cos\left(\vec{K}\cdot\frac{a}{2}(\hat{x}+\hat{z})\right) - 2\alpha \cos\left(\vec{K}\cdot\frac{a}{2}(-\hat{x}+\hat{z})\right) \\ &- 2\beta \cos\left(\vec{K}\cdot\frac{a}{2}(\hat{y}+\hat{z})\right) - 2\beta \cos\left(\vec{K}\cdot\frac{a}{2}(-\hat{y}+\hat{z})\right) + 8\alpha + 4\beta\Big]. \end{aligned} \tag{3.81}$$

Simplifying Equation 3.81 gives

$$\begin{aligned} D_{xx}(\vec{K}) = \frac{4}{M}\Big[&\alpha \sin^2\left(\vec{K}\cdot\frac{a}{4}(\hat{x}+\hat{y})\right) + \alpha \sin^2\left(\vec{K}\cdot\frac{a}{4}(-\hat{x}+\hat{y})\right) \\ &+ \alpha \sin^2\left(\vec{K}\cdot\frac{a}{4}(\hat{x}+\hat{z})\right) + \alpha \sin^2\left(\vec{K}\cdot\frac{a}{4}(-\hat{x}+\hat{z})\right) \\ &+ \beta \sin^2\left(\vec{K}\cdot\frac{a}{4}(\hat{y}+\hat{z})\right) + \beta \sin^2\left(\vec{K}\cdot\frac{a}{4}(-\hat{y}+\hat{z})\right)\Big]. \end{aligned} \tag{3.82}$$

The other diagonal elements follow quickly,

$$\begin{aligned} D_{yy}(\vec{K}) = \frac{4}{M}\Big[&\alpha \sin^2\left(\vec{K}\cdot\frac{a}{4}(\hat{x}+\hat{y})\right) + \alpha \sin^2\left(\vec{K}\cdot\frac{a}{4}(-\hat{x}+\hat{y})\right) \\ &+ \beta \sin^2\left(\vec{K}\cdot\frac{a}{4}(\hat{x}+\hat{z})\right) + \beta \sin^2\left(\vec{K}\cdot\frac{a}{4}(-\hat{x}+\hat{z})\right) \\ &+ \alpha \sin^2\left(\vec{K}\cdot\frac{a}{4}(\hat{y}+\hat{z})\right) + \alpha \sin^2\left(\vec{K}\cdot\frac{a}{4}(-\hat{y}+\hat{z})\right)\Big], \end{aligned} \tag{3.83}$$

and

$$\begin{aligned} D_{zz}(\vec{K}) = \frac{4}{M}\Big[&\beta \sin^2\left(\vec{K}\cdot\frac{a}{4}(\hat{x}+\hat{y})\right) + \beta \sin^2\left(\vec{K}\cdot\frac{a}{4}(-\hat{x}+\hat{y})\right) \\ &+ \alpha \sin^2\left(\vec{K}\cdot\frac{a}{4}(\hat{x}+\hat{z})\right) + \alpha \sin^2\left(\vec{K}\cdot\frac{a}{4}(-\hat{x}+\hat{z})\right) \\ &+ \alpha \sin^2\left(\vec{K}\cdot\frac{a}{4}(\hat{y}+\hat{z})\right) + \alpha \sin^2\left(\vec{K}\cdot\frac{a}{4}(-\hat{y}+\hat{z})\right)\Big]. \end{aligned} \tag{3.84}$$

The off-diagonal elements are

$$D_{xy}(\vec{K}) = \frac{1}{M}\Big[- 2\gamma \cos\left(\vec{K}\cdot\frac{a}{2}(\hat{x}+\hat{y})\right) + 2\gamma \cos\left(\vec{K}\cdot\frac{a}{2}(-\hat{x}+\hat{y})\right)\Big], \tag{3.85}$$

which may be rewritten as

$$D_{xy}(\vec{K}) = \frac{4\gamma}{M}\Big[\sin\left(\vec{K}\cdot\left(\frac{a}{2}\right)\hat{x}\right) \sin\left(\vec{K}\cdot\left(\frac{a}{2}\right)\hat{y}\right)\Big]. \tag{3.86}$$

The other two off-diagonal elements follow immediately

$$D_{xz}(\vec{K}) = \frac{4\gamma}{M}\left[\sin\left(\vec{K}\cdot\left(\frac{a}{2}\right)\hat{x}\right)\sin\left(\vec{K}\cdot\left(\frac{a}{2}\right)\hat{z}\right)\right]. \tag{3.87}$$

$$D_{yz}(\vec{K}) = \frac{4\gamma}{M}\left[\sin\left(\vec{K}\cdot\left(\frac{a}{2}\right)\hat{y}\right)\sin\left(\vec{K}\cdot\left(\frac{a}{2}\right)\hat{z}\right)\right]. \tag{3.88}$$

The dynamical matrix is symmetrical.

Finally the stage is set to calculate the dispersion curves for a specific fcc case. The ω *vs.* $\vec{K}$ curve will be calculated for $\vec{K}$ oriented along various high-symmetry directions in the fcc lattice. The terms Γ, X, K, and L refer to high-symmetry points in the Brillouin zone. It is not necessary to be familiar with Brillouin zones to comprehend the following discussion. The coordinates of these points in the Brillouin zone are as follows: $\Gamma \rightarrow (2\pi/a)(0,0,0)$; $X \rightarrow (2\pi/a)(1,0,0)$; $K \rightarrow (2\pi/a)(3/4,3/4,0)$; and, $L \rightarrow (2\pi/a)(1/2,1/2,1/2)$. Γ is at the Brillouin zone center, while the others lie on the boundary of the first Brillouin zone. As an example a wave vector $0 \leq \vec{K} \leq (2\pi/a)(1/2,1/2,1/2)$, indicates a wave in the [111] direction with $\vec{K}$ varying between 0 and the L point. In what follows the [100], [110], and [111] directions will be explored. For the present case Equation 3.69 becomes,

$$\begin{pmatrix} D_{xx}(\vec{K}) - \omega^2 & D_{xy}(\vec{K}) & D_{xz}(\vec{K}) \\ D_{xy}(\vec{K}) & D_{yy}(\vec{K}) - \omega^2 & D_{yz}(\vec{K}) \\ D_{xz}(\vec{K}) & D_{yz}(\vec{K}) & D_{zz}(\vec{K}) - \omega^2 \end{pmatrix}\begin{pmatrix} e_x \\ e_y \\ e_z \end{pmatrix} = 0. \tag{3.89}$$

The [100] direction will be explored first. It is convenient to set $\vec{K} = (2\pi/a)(\zeta,0,0)$ with $0 \leq \zeta \leq 1$. Using this value of $\vec{K}$ in Equations 3.82 through 3.88 results in

$$D_{xx}(\zeta) = \frac{16\alpha}{M}\sin^2(\zeta\pi/2), \tag{3.90}$$

$$D_{yy}(\zeta) = D_{zz}(\zeta) = \frac{(8\alpha + 8\beta)}{M}\sin^2(\zeta\pi/2) \tag{3.91}$$

and

$$D_{xy}(\zeta) = D_{xz}(\zeta) = D_{yz}(\zeta) = 0. \tag{3.92}$$

Because the matrix is diagonal, the solutions are obvious,

$$\omega_1 = \sqrt{\frac{16\alpha}{M}\sin^2(\zeta\pi/2)} \tag{3.93}$$

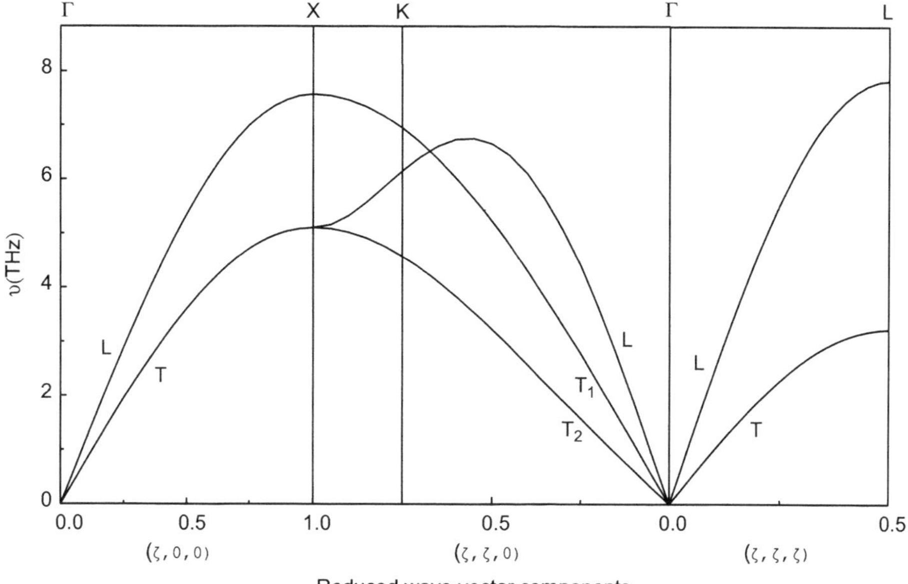

Figure 3.7 Phonon dispersion curve for copper, calculated using only nearest-neighbor force constants. The vertical axes is $\nu = \omega/2\pi$. The Cu force constants were taken from Ref. [51]. The reduced wave vector components are the components of $\vec{K}$ divided by $2\pi/a$. The symmetry points Γ, X, K, and L shown at the top of the graph are discussed in the text. The L and T within the figure indicate longitudinal and transverse waves.

a longitudinal wave with particle motion in the $\hat{x}_1$ direction – the same as the 1-D solution , Equation 3.46 – and

$$\omega_2 = \omega_3 = \sqrt{\frac{8\alpha + 8\beta}{M} \sin^2(\zeta\pi/2)}, \tag{3.94}$$

two transverse waves with the same frequencies (degenerate solutions) polarized in the $\hat{x}_2$ and $\hat{x}_3$ directions respectively. These results are illustrated in the Γ to X panel of Figure 3.7.

The solution for the [110] direction proceeds in a similar manner, although the results are more complicated. In this case $0 \leq \vec{K} \leq (2\pi/a)(\zeta, \zeta, 0)$. Again, $0 \leq \zeta \leq 1$. Then,

$$D_{xx}(\zeta) = D_{yy}(\zeta) = \frac{4\alpha}{M} \sin^2(\zeta\pi) + \frac{8\alpha + 8\beta}{M} \sin^2(\zeta\pi/2), \tag{3.95}$$

$$D_{zz}(\zeta) = \frac{4\beta}{M} \sin^2(\zeta\pi) + \frac{16\alpha}{M} \sin^2(\zeta\pi/2), \tag{3.96}$$

and

$$D_{xy}(\zeta) = \frac{4\gamma}{M}\sin^2(\zeta\pi). \tag{3.97}$$

These equations can be solved for an analytical expression for the frequencies, but the results are somewhat complicated and do not seem to provide a great deal of insight. A different approach will be discussed below.

But next, propagation in the [111] direction will be examined. For this example $0 \le \vec{K} \le (2\pi/a)(\zeta,\zeta,\zeta)$, with $0 \le \zeta \le 1/2$. The matrix elements are

$$D_{xx}(\zeta) = D_{yy}(\zeta) = D_{zz}(\zeta) = \frac{8\alpha + 4\beta}{M}\sin^2(\zeta\pi), \tag{3.98}$$

and

$$D_{xy}(\zeta) = D_{xz}(\zeta) = D_{yz}(\zeta) = \frac{4\gamma}{M}\sin^2(\zeta\pi). \tag{3.99}$$

At some point it is better to abandon more and more complicated algebra and use a computer. Numerical results are especially easy to obtain once the matrix elements are found. One simply forms the matrix

$$\begin{pmatrix} D_{xx}(\vec{K}) & D_{xy}(\vec{K}) & D_{xz}(\vec{K}) \\ D_{xy}(\vec{K}) & D_{yy}(\vec{K}) & D_{yz}(\vec{K}) \\ D_{xz}(\vec{K}) & D_{yz}(\vec{K}) & D_{zz}(\vec{K}) \end{pmatrix}, \tag{3.100}$$

provides the numerical data, and uses one of the several commercially available software programs to find the eigenvalue and eigenvectors. Just such an approach was used to produce Figure 3.7. For that figure the nearest-neighbor force constants for Cu were used [51]: $\alpha = 14.9$; $\gamma = 17.6$ $\beta = -1.34$ all in units of J/m^2. Neutron scattering is usually used to experimentally determine dispersion curves such as those of Figure 3.7. In practice, many different units seem to be in use for the vertical axes. For the present case the frequency in Hertz, not radians per second, was used, $\nu = \omega/2\pi$. Although only first-neighbor forces were used, Figure 3.7 appears to be a good approximation to the actual data [52]. This result is consistent with the fact that the second, and higher, neighbor force constants are considerably smaller than the first-neighbor constants [52].

Comparison of 1D case, Figure 3.2, with the 3D case, Figure 3.7, shows similarities and differences. The similarities are in the general shapes of the dispersion curves; the difference is that where there is only one wave for each $\vec{K}$ value in 1D, there are three in the 3D case, one longitudinal and two transverse. (For a general direction in 3D the waves may be quasi-longitudinal and quasi-transverse.)

It is also interesting to compare the present results with those of Section 3.1.1. For example, both cases show that the transverse waves are degenerate in the [100] and [111] directions, but not in the [110] direction.

New features emerge when there are two atoms per Bravais lattice point (two atoms in the basis). Figure 3.4 provides a guide. For two atoms in the basis in the 3D case, there are three optic modes and three acoustic modes for each $\vec{K}$ value [53]. For N atoms in the basis there are three acoustic modes and $3(N-1)$ optic modes.

Force Constants and Elastic Constants

Acoustic waves in the long-wavelength limit were explored in Section 3.1. In Section 3.1.1 the phase velocity $v = \omega/K$ was found in terms of the elastic constants for a number of cases. Figure 3.7 indicates that for $Ka << 1$ the phase velocity is constant and can be found in terms of the lattice-dynamical force constants. Thus, it should be possible to get an approximate relation between the elastic constants and the force constants. The [110] direction for fcc symmetry will be investigated. Section 3.1.1 gives the relevant classical long-wavelength solutions for v in terms of the C_{ij}.

To get the results from the lattice-dynamical approach, Equations 3.95–3.97 will be approximated for $Ka << 1$. For the [110] direction the magnitude of $\vec{K}$ is $K = (2\pi/a)\sqrt{\zeta^2+\zeta^2} = (2\sqrt{2}/a)\pi\zeta$. Solving for ζ gives $\zeta = aK/(2\sqrt{2}\pi)$. Using $\sin(x) \approx x$ for $x << 1$ in Equations 3.95–3.97 leads to the dynamical matrix for $Ka << 1$,

$$\frac{K^2a^2}{M}\begin{pmatrix} \frac{3\alpha+\beta}{4} & \frac{\gamma}{2} & 0 \\ \frac{\gamma}{2} & \frac{3\alpha+\beta}{4} & 0 \\ 0 & 0 & \frac{\alpha+\beta}{2} \end{pmatrix}. \tag{3.101}$$

Work in previous sections has shown that the eigenvalues of this matrix give the squares of ω and the eigenvectors give the polarizations. The first eigenvalue gives,

$$\omega_1^2 = \frac{K^2a^2}{M}\left(\frac{3\alpha+\beta+2\gamma}{4}\right) \tag{3.102}$$

and $e_x = e_y$. In other terms,

$$v_l = \frac{\omega_1}{K} = a\sqrt{\frac{3\alpha+\beta+2\gamma}{4M}}, \tag{3.103}$$

for a longitudinal wave traveling in the [110] direction. The second eigenvalue gives,

$$v_{t1} = \frac{\omega_2}{K} = a\sqrt{\frac{3\alpha+\beta-2\gamma}{4M}}, \tag{3.104}$$

Table 3.2 *Cu elastic constants (GPa).*

	Equations 3.106	Reference [55]
c_{11}	165	168
c_{12}	120	121
c_{44}	75	75

and $e_x = -e_y$ for a transverse wave polarized in the $[1\bar{1}0]$ direction. The third eigenvalue gives

$$v_{t2} = \frac{\omega_3}{K} = a\sqrt{\frac{\alpha + \beta}{2M}}, \tag{3.105}$$

with $e_x = e_y = 0$, a transverse wave polarized in the [001] direction.

The results from the Christoffel equations involve the density, ρ, which for the fcc lattice is $4M/a^3$. Using this result for ρ and equating corresponding velocities from Equations 3.103, 3.104, and 3.105 to those of Equations 3.19, 3.22, and 3.24 results in,

$$c_{11} = \frac{4\alpha}{a} \qquad c_{44} = \frac{2(\alpha + \beta)}{a} \qquad c_{12} = \frac{2(2\gamma - \alpha - \beta)}{a}. \tag{3.106}$$

Because only first-neighbor force constants are used, it is not reasonable to expect great accuracy; however, Equation 3.106 agrees rather well (perhaps fortuitously) with experimental results for Cu. Table 3.2 shows a comparison. The calculated results used the force constants mentioned earlier, $\alpha = 14.9$; $\gamma = 17.6$; $\beta = -1.34$ all in units of J/m^2, [51], and a Cu lattice constant of 0.3615 nm [54]. The very close agreement between the two sets of c_{ij} for Cu seems to be an exception. Several percent disagreement seems to be more typical.

3.2.4 Normal Modes and Quantization of Lattice Vibrations

In Section 3.2.3, the displacements of the atoms were found (Equation 3.64) to be $u_\alpha(n, l, t) = A_\alpha(n) \exp[i(\vec{K} \cdot \vec{R}_l - \omega t)]$. The $u_\alpha(n, l, t)$ are the normal modes of vibrations for the atoms.[11] The atomic motion is described in terms of waves, rather than independent atomic vibrations. Much of Section 3.2.3 was devoted to finding the relation between ω and $\vec{K}$, the dispersion relation. However, the solution is not complete until boundary conditions are imposed. The imposition of boundary conditions for a finite sample results in only discrete values of $\vec{K}$ being allowed, as will now be shown.

[11] The present treatment is based on the harmonic approximation. Only terms to second-order in the interatomic potential are included, Equation 3.59. When higher-order terms are included, the $u_\alpha(n, l, t)$ are not the exact normal modes.

A reasonable set of boundary conditions might be to require the displacement, $u_\alpha(n,l,t)$, to be held fixed, possibly zero, on the boundaries of the specimen; however, fixed boundaries are not the most mathematically convenient. It is usual to use periodic boundary conditions in which the solutions repeat themselves periodically [39, 40, 41, 42]. Suppose the specimen is chosen as a rectangular parallelepiped of lengths L_1, L_2, L_3 on the sides, with the edges oriented along the x_1, x_2, x_3 axes respectively. Then, periodic boundary conditions are, $u(x_1,x_2,x_3) = u(x_1 + L_1, x_2, x_3)$, with similar requirements for the x_2 and x_3 directions.[12] In terms of the present notation the periodic boundary conditions are expressed as

$$\begin{aligned} A_\alpha(n)\exp[i(\vec{K}\cdot\vec{R}_l - \omega t) &= A_\alpha(n)\exp[i(\vec{K}\cdot(\vec{R}_l + L_1\hat{\mathrm{x}}_1) - \omega t) \\ A_\alpha(n)\exp[i(\vec{K}\cdot\vec{R}_l - \omega t) &= A_\alpha(n)\exp[i(\vec{K}\cdot(\vec{R}_l + L_2\hat{\mathrm{x}}_2) - \omega t) \\ A_\alpha(n)\exp[i(\vec{K}\cdot\vec{R}_l - \omega t) &= A_\alpha(n)\exp[i(\vec{K}\cdot(\vec{R}_l + L_3\hat{\mathrm{x}}_3) - \omega t). \end{aligned} \tag{3.107}$$

These conditions are satisfied if

$$K_1 = n_1\frac{2\pi}{L_1} \qquad K_2 = n_2\frac{2\pi}{L_2} \qquad K_3 = n_3\frac{2\pi}{L_3}, \tag{3.108}$$

where n_1, n_2, and n_3 are integers, including zero and negative values. Thus, the values of $\vec{K}$, *e.g.*, in Figure 3.7, are not continuous, but discrete, although for macroscopic specimens the values are extremely close together. Equations 3.108 are needed primarily for finding the density of the allowed values of $\vec{K}$.

The situation is most easily illustrated in two dimensions. The case for $L_2 = 2L_1$ shown in Figure 3.8, illustrates the K points generated as n_1 and n_2 range through a set of integer values around zero. The highlighted K-space cell has four K points at the corners, but these are each shared four ways, resulting in one K point for a K-space area of $(2\pi)^2/(L_1L_2) = 2\pi^2/A$ where A is the real-space area of a rectangular specimen of dimensions L_1, L_2. The number of K points per unit area of K-space is $A/(2\pi)^2$.

It is easy to generalize the two-dimensional result to three dimensions. In the 3D case, K-space is filled with rectangular parallelepiped cells of volume $(2\pi)^3/(L_1L_2L_3)$ with an allowed K point at each of the eight corners. But, the corner points are shared among eight parallelepipeds, so there is one K point per unit volume of K-space. Then,

$$K\text{-space density of points} = V/(2\pi)^3 \tag{3.109}$$

with $V = L_1L_2L_3$, the real-space volume of the parallelepiped. Equation 3.109 is of major importance, and will be used several times in what follows.

[12] For $L >>$ the lattice constant, the different types of boundary conditions give the same result to a good approximation. See [40, 41] for more detailed discussions.

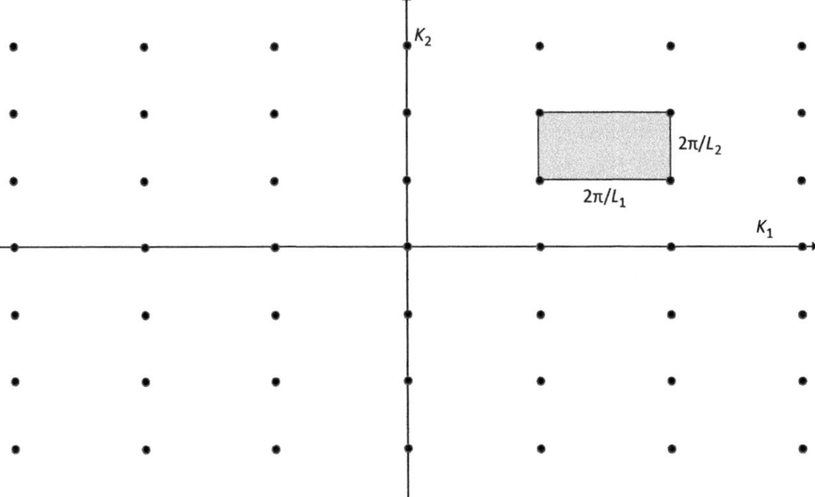

Figure 3.8 Allowed K points for a two-dimensional case. This illustration is for the case $L_2 = 2L_1$. There are four points at the corners of the highlighted cell, but these are shared four ways. Thus, there is one K point for a K-space area of $(2\pi)^2/(L_1L_2) = (2\pi)^2/A$.

To summarize briefly, the normal modes for the lattice vibrations are given by $u_\alpha(n,l,t) = A_\alpha(n)\exp[i(\vec{K}\cdot\vec{R}_l - \omega t)]$ – Equation 3.64 – with the allowed $\vec{K}$ values given by Equation 3.108. All unique solutions are given by $\vec{K}$ values within the first Brillouin zone.

As might have been expected from the fact that the restoring forces involved in the harmonic approach are proportional to the displacement of atoms from equilibrium positions, the energy of each normal mode is just that of a harmonic oscillator [39, 40]. For a simple harmonic oscillator the possible energies, E_s, are

$$E_s = (s + 1/2)\hbar\omega, \tag{3.110}$$

where $\hbar$ is Planck's constant, ω is the angular frequency of the oscillator, and $s = 0, 1, 2, 3...$ The *thermal average* energy in mode (K, p) is[13]

$$u_{K,p} = \left(\langle s_{K,p}\rangle + \frac{1}{2}\right)\hbar\omega_{K,p}, \tag{3.111}$$

where $\langle s_{K,p}\rangle$, the thermal average number of quantum excitations in mode (K, p), is given by the Planck distribution function [56, 57]

$$\langle s_{K,p}\rangle = \frac{1}{\exp\left(\hbar\omega_{K,p}/k_BT\right) - 1}. \tag{3.112}$$

[13] There are three polarizations, p, for each K value.

The total lattice vibrational energy is

$$E_{lattice} = \sum_K \sum_p u_{K,p}. \tag{3.113}$$

In principle, the lattice contribution to the heat capacity can be calculated from Equation 3.113.

Although it is accurate to call $\langle s_{K,p} \rangle$ the average number of quantum excitations in mode (K, p), it is more convenient, and common practice, to speak of the number of phonons of wave vector $\vec{K}$ and polarization p in the specimen [40]. The terminology is quite analogous to the use of photons to describe electromagnetic waves.

3.3 Debye Theory of Solids

3.3.1 Introduction

The development of the current understanding of the specific heat of solids illustrates nicely the overall progress in the theory of the solid state over the past century or so. Early attempts at a theoretical description of the specific heat of solids were spectacular failures. Success awaited the application of quantum mechanics to the problem, and an understanding of the normal modes of vibration of crystal lattices.

The early classical approach treated atoms as independent harmonic oscillators. For N atoms there are $3N$ oscillators corresponding to vibrations in the three x_i directions. Classical statistical mechanics [40, 42, 58] gives the average thermal energy of each harmonic oscillator as $k_B T$, resulting in a total vibrational energy of $U = 3Nk_B T$. The resulting specific heat $c_v = \partial U / \partial T = 3Nk_B$ was in reasonably good agreement with experiments for high temperatures, but failed miserably at low temperatures where the experimental results approached zero.

In 1907 Einstein made a famous contribution to the subject [59]. As in the classical approach, he treated the atoms as $3N$ independent harmonic oscillators, but the oscillators were taken to obey quantum mechanics with the energies given by Equations 3.111 and 3.112. All oscillators were assumed to have the same frequency, now called ω_E. The specific heat resulting from this model is

$$c_v = 3Nk_B \left(\frac{\hbar\omega_E}{k_B T} \right)^2 \frac{\exp\left(\frac{\hbar\omega_E}{k_B T}\right)}{\left(\exp\left(\frac{\hbar\omega_E}{k_B T}\right) - 1\right)^2}. \tag{3.114}$$

Equation 3.114 produced a dramatic improvement in the agreement between theory and experiment, going to zero at $T = 0$ and approaching $3Nk_B$ at high temperatures. Although the Einstein model was a great improvement over the classical model, there were still serious disagreements with experimental results.[14] In 1912 Debye made another big step forward [60]. However, before discussing Debye's contribution it is useful to develop some ideas of the density of states, $D(\omega)$, where $D(\omega)d\omega$ is the number of states (vibrational modes) between ω and $\omega + d\omega$. The need for $D(\omega)$ arises because a fundamental quantity is the total vibrational energy, which depends on ω. Equation 3.111 shows the contribution from each mode, $(\vec{K}, p)$. Because the $\vec{K}$ states are uniformly distributed in $\vec{K}$ space, it is trivial to count them, but getting the results in terms of ω can be more difficult.

Recalling that in general there are three normal modes for each $\vec{K}$ value – *i.e.*, the ω *vs.* $\vec{K}$ curve has three branches – consider for the moment *only one branch* of the curve, then [39, 40]

$$D(\omega)d\omega = \frac{\int_{shell} d^3\vec{K}}{(2\pi)^3/V}. \tag{3.115}$$

The integral gives a volume in $\vec{K}$ space and the denominator gives the volume per $\vec{K}$ point, Equation 3.109, so that the right-hand side gives the number of $\vec{K}$ points in the volume covered by the integral. To make the connection with ω, the shell is bounded by one surface on which ω is constant and by another surface on which $\omega + d\omega$ is constant. Figure 3.7 shows that these surfaces could be a bit complicated. The left-hand side of Equation 3.115 gives the number of states in terms of ω which, by definition, is equal to the number of states on the right-hand side of the equation.

Debye advanced Einstein's contribution by taking the normal modes of vibration to be long-wavelength sound waves, not independent atomic vibrations. For these modes, the frequency depends linearly on the wave vector, as shown in Figure 3.7. Each of these modes has the harmonic oscillator energy spectrum, Equation 3.111. Initially the specimen will be taken as isotropic. In that case, the constant ω surfaces are spheres. The integral, for a shell thickness of $d\vec{K}$, is $4\pi K^2 dK$, resulting in

$$D(\omega)d\omega = \frac{V}{2\pi^2} K^2 dK \tag{3.116}$$

for *each branch* of the dispersion curve.

An oversimplified example will be discussed first, which will then be extended. For the linear dispersion relation take (Equation 3.7)

$$\omega = vK \tag{3.117}$$

where v is the phase velocity. In an elastic 3D solid there are *three* waves for each distinct K value, one longitudinal (or quasi-longitudinal), and two transverse (or

[14] See p. 365 of Ref. [42].

quasi-transverse). For the moment, the unrealistic assumption is made that all three waves have the same velocity. Then, using Equations 3.116 and 3.117

$$D(\omega) = \frac{3V\omega^2}{2\pi^2 v^3} \tag{3.118}$$

where the factor 3 is for the three modes for each $\vec{K}$. This result can be improved by using two relations such as Equation 3.117, one for longitudinal waves with velocity v_l and one for the two transverse modes with velocities v_t. Then

$$D(\omega) = \frac{3V\omega^2}{2\pi^2 v_o^3}. \tag{3.119}$$

with

$$\frac{3}{v_o^3} = \left(\frac{1}{v_l^3} + \frac{2}{v_t^3}\right). \tag{3.120}$$

Debye realized that the discrete nature of the lattice prevented Equation 3.117 from being extended to arbitrarily high frequencies. He selected a cut-off frequency, now commonly called ω_D, by requiring the total number of modes to equal $3N$,

$$3N = \int_0^{\omega_D} D(\omega) d\omega = \frac{V\omega_D^3}{2\pi^2 v_o^3}. \tag{3.121}$$

Solving Equation 3.121 for the Debye frequency gives,

$$\omega_D = \left(6\pi^2 \frac{N}{V}\right)^{1/3} v_o, \tag{3.122}$$

a characteristic frequency of the material.

For a material with only one atom per basis, N/V is simply the number of atoms per volume. For a material with a polyatomic basis, there are two reasonable choices for N/V.[15] One choice is to continue to take N/V as the number of atoms per volume, so that all vibrations are included in the Debye model. *This is the choice that is usually made.* A second choice is to treat the acoustic modes and the optic modes differently. The optic modes are treated within the Einstein model, while the acoustic modes are treated within the Debye model. In this second case, N/V for the Debye contribution is just the number of primitive unit cells per volume. The second choice has frequently been chosen for the case of light interstitials in metals [61, 62, 63], where the optic modes are at relatively high frequencies and usually well-separated from the acoustic modes.

[15] See p. 462 of Ref. [40] for an excellent discussion of this point.

3.3.2 Lattice Thermal Energy and Specific Heat

The stage is set to calculate the specific heat with the Debye model. The lattice vibrational energy is

$$U = \int_0^{\omega_D} u(\omega)D(\omega)d\omega \tag{3.123}$$

where, from Equations 3.111 and 3.112

$$u(\omega) = \left(\frac{1}{2} + \frac{1}{e^{(\hbar\omega/k_BT)} - 1}\right)\hbar\omega, \tag{3.124}$$

and $D(\omega)$ is given by Equation 3.119. Substituting,

$$U = U_o + \frac{3V\hbar}{2\pi^2 v_o^3}\int_0^{\omega_D} \frac{\omega^3 d\omega}{e^{(\hbar\omega/k_BT)} - 1}, \tag{3.125}$$

where U_o represents the temperature independent zero-point energy term, and does not contribute to the specfic heat.[16] The specific heat is found by differentiating U with respect to T,

$$c_v = \frac{3V\hbar^2}{2\pi^2 v_o^3 k_B T^2}\int_o^{\omega_D} \frac{\omega^4 e^{(\hbar\omega/k_BT)}d\omega}{\left(e^{(\hbar\omega/k_BT)} - 1\right)^2}. \tag{3.126}$$

Equations 3.125 and 3.126 are both put in a form convenient for integration by defining

$$x \equiv \hbar\omega/k_BT \qquad x_D \equiv \hbar\omega_D/k_BT = \theta_D/T, \tag{3.127}$$

where θ_D is defined as the Debye temperature, $\theta_D = \hbar\omega_D/k_B$. Using Equation 3.122 the Debye temperature may be written

$$\theta_D = \frac{\hbar}{k_B}\left[6\pi^2\frac{N}{V}\right]^{1/3} v_o. \tag{3.128}$$

The equations for U and c_v take the form,

$$U = U_o + \frac{3Vk_B^4T^4}{2\pi^2 v_o^3\hbar^3}\int_0^{x_D} \frac{x^3}{e^x - 1}dx, \tag{3.129}$$

and

$$c_v = 9Nk_B\left[\frac{T}{\theta_D}\right]^3 \int_0^{x_D} \frac{x^4 e^x}{(e^x - 1)^2}dx. \tag{3.130}$$

[16] This term takes a very simple form [42], $U_o = \frac{9}{8}Nk_B\theta_D$, obtained from $\int_o^{\omega_D} \frac{1}{2}\hbar\omega D(\omega)d\omega$.

This is a famous result. It contains only one material parameter, θ_D. It is remarkably simple and successful. (See, *e.g.* figure 11.5 of Ref. [42] and figure 3.13 of Ref. [64].) Note that x_D depends on the temperature. Given θ_D, or finding it from fitting data, the specific heat can be calculated over an extended temperature range.

It is interesting to examine the high and low temperature limits. It is convenient to look at Equation 3.129. At low temperatures, $T << \theta_D$ and $x_D = \theta_D/T$ becomes large. Due to the exponential in the denominator, there is little contribution to the integrand of Equation 3.125 for large x_D and there is little error in letting the upper limit go to infinity. This integral is discussed in various books on solid state physics [39, 42]. The limiting value of the integrand is $\pi^4/15$. Then,

$$U \simeq U_o + \frac{3}{5}\pi^4 N k_B \frac{T^4}{\theta_D^3}. \tag{3.131}$$

The thermal vibrational energy varies as T^4 in the low-temperature limit. The specific heat follows immediately,

$$c_v = \frac{12}{5}\pi^4 N k_B \left(\frac{T}{\theta_D}\right)^3, \tag{3.132}$$

the well-known T^3 law for the low-temperature specific heat.

Next, the high-temperature limit is considered. The expression for $u(\omega)$ can be simplified in this limit, which is $k_BT >> \hbar\omega$ for all frequencies.

$$\begin{aligned} u(\omega) &= \hbar\omega\left(\frac{1}{2} + \frac{1}{\exp\frac{\hbar\omega}{k_BT} - 1}\right) \simeq \hbar\omega\left(\frac{1}{2} + \frac{1}{\frac{\hbar\omega}{k_BT} + \frac{1}{2}\left(\frac{\hbar\omega}{k_BT}\right)^2 \cdots}\right) \\ &= \hbar\omega\left(\frac{1}{2} + \frac{1}{\left(\frac{\hbar\omega}{k_BT}\right)\left(1 + \frac{1}{2}\left(\frac{\hbar\omega}{k_BT}\right)\right)}\right) \simeq \hbar\omega\left(\frac{1}{2} + \frac{k_BT}{\hbar\omega}\left(1 - \frac{1}{2}\frac{\hbar\omega}{k_BT}\right)\right) = k_BT. \end{aligned} \tag{3.133}$$

Notice how the factor $\left(\frac{1}{2}\right)\left(\frac{\hbar\omega}{k_BT}\right)^2$ in the expansion of the exponential exactly cancels the zero-point motion contribution [40]. The total thermal vibrational energy in the high temperature limit is,

$$U = 3Nk_BT \tag{3.134}$$

and the specific heat is

$$c_v = 3Nk_B, \tag{3.135}$$

the classical result for the high-temperature regime. As is usually the case, the quantum results go over to the classical results in the appropriate limit.

It is now possible to understand why the Debye model works so well, even though the basic assumption of Equation 3.117 is rather drastic. Equation 3.124 shows that as k_BT becomes less than $\hbar\omega$ for a particular mode, the contribution of that mode to the thermal energy begins to decrease exponentially as $\approx e^{(-\hbar\omega/k_BT)}$. Thus, at low temperature only the low-frequency modes make an appreciable contribution to the thermal energy, just the modes for which the Debye approximation is valid. On the other hand, as Equation 3.133 shows, at high temperatures all modes make a contribution of k_BT to the thermal energy, no matter if the dispersion relation is that of Equation 3.117 or the more realistic dispersion relation of Figure 3.7. It is perhaps not surprising that the Debye model, constrained to work at low temperatures and high temperatures, also fits experimental results reasonably well in the intermediate range.

3.3.3 Debye Temperature from Elastic Constants

The discussions of the Debye theory in the previous Section were based on an isotropic solid. For anisotropic solids a straight-forward generalization of Equation 3.120 is [65, 66],

$$\frac{3}{v_o^3} = \frac{\int \left(\frac{1}{v_l^3(\theta,\phi)} + \frac{1}{v_{t1}^3(\theta,\phi)} + \frac{1}{v_{t2}^3(\theta,\phi)}\right)}{\int \sin(\theta)d\theta d\phi} \sin(\theta)d\theta d\phi \tag{3.136}$$

where the integral is over the solid angle Ω with $d\Omega = \sin(\theta)d\theta d\phi$. The equation is expressed in terms of the usual angles of spherical coordinates, θ and ϕ. The direction cosines of Equation 3.8 can be expressed in terms of θ and ϕ,

$$n_1(\theta,\phi) = \sin(\theta)\cos(\phi),\ n_2(\theta,\phi) = \sin(\theta)\sin(\phi),\ n_3(\theta,\phi) = \cos(\theta). \tag{3.137}$$

The phase velocities of Equation 3.136, apart from a factor ρ, may be found from the eigenvalues of the Christoffel matrix (Equation 3.8),

$$\Gamma(\theta,\phi) = \begin{pmatrix} \Gamma_{11}(\theta,\phi) & \Gamma_{12}(\theta,\phi) & \Gamma_{13}(\theta,\phi) \\ \Gamma_{12}(\theta,\phi) & \Gamma_{22}(\theta,\phi) & \Gamma_{23}(\theta,\phi) \\ \Gamma_{13}(\theta,\phi) & \Gamma_{23}(\theta,\phi) & \Gamma_{33}(\theta,\phi) \end{pmatrix}. \tag{3.138}$$

Equations 3.137 are used to change from direction cosines to spherical coordinates. The eigenvalues of Equation 3.138 are $\rho v_l(\theta,\phi)$, $\rho v_{t1}(\theta,\phi)$ and $\rho v_{t2}(\theta,\phi)$.

For practical computations it is convenient to change Equation 3.136 from an integral to a sum,

$$\frac{3}{v_o^3} = \frac{\sum_i \sum_j \left[\left(\frac{1}{v_l^3(\theta_i,\phi_j)} + \frac{1}{v_{t1}^3(\theta_i,\phi_j)} + \frac{1}{v_{t2}^3(\theta_i,\phi_j)}\right) \sin(\theta_i)\Delta\theta_i\Delta\phi_j\right]}{\sum_i \sum_j \sin(\theta_i)\Delta\theta_i\Delta\phi_j}. \tag{3.139}$$

It makes sense to take $\Delta\theta_i = \Delta\theta$ and $\Delta\phi_j = \Delta\phi$ as constants. These constants can be taken outside the summations with the simplification,

$$\frac{3}{v_o^3} = \frac{\sum_i \sum_j \left(\frac{1}{v_l^3(\theta_i,\phi_j)} + \frac{1}{v_{t1}^3(\theta_i,\phi_j)} + \frac{1}{v_{t2}^3(\theta_i,\phi_j)} \right)}{\sum_i \sum_j \sin(\theta_i)} \sin(\theta_i). \tag{3.140}$$

It is important to keep both sums in the denominator of Equation 3.140 for normalization. The angles θ_i and ϕ_j can be varied to cover the whole 4π solid angle, although, by symmetry arguments a fraction of 4π is often sufficient. Obviously, a larger number of directions should produce a more accurate value of Θ_D.

For cubic symmetry the Christoffel matrix elements are, Equation 3.12,

$$\begin{aligned}
\Gamma_{11}(\theta,\phi) &= n_1^2(\theta,\phi)c_{11} + (n_2^2(\theta,\phi) + n_3^2(\theta,\phi))c_{44} \\
\Gamma_{12}(\theta,\phi) &= n_1(\theta,\phi)n_2(\theta,\phi)(c_{12} + c_{44}) \\
\Gamma_{13}(\theta,\phi) &= n_1(\theta,\phi)n_3(\theta,\phi)(c_{12} + c_{44}) \\
\Gamma_{22}(\theta,\phi) &= n_2^2(\theta,\phi)c_{11} + (n_1^2(\theta,\phi) + n_3^2(\theta,\phi))c_{44} \\
\Gamma_{23}(\theta,\phi) &= n_2(\theta,\phi)n_3(\theta,\phi)(c_{12} + c_{44}) \\
\Gamma_{33}(\theta,\phi) &= n_3^2(\theta,\phi)c_{11} + (n_1^2(\theta,\phi) + n_2^2(\theta,\phi))c_{44}.
\end{aligned} \tag{3.141}$$

In the present case – cubic symmetry – it is sufficient to vary θ from 0 to $\pi/2$ and ϕ from 0 to $\pi/4$, which covers only $1/16$ of the full 4π solid angle. Step sizes of $\pi/1000$, corresponding to 125,000 different directions in the $1/16$ part of the full solid angle, do not pose a problem with present-day portable computers.

To summarize, given the elastic constants and ρ, the velocities for any direction can be obtained from the eigenvalues of $\Gamma(\theta,\theta)$. Equation 3.140 gives $3/v_o^3$, which is then used with Equation 3.128 to obtain the Debye temperature for an anisotropic material. This procedure can be performed with one of the several commerically available software programs [67].

Within the Debye model, the phonon contribution to several thermodynamic quantities can be determined once the Debye temperature is known. The internal energy, U, and the specific heat, c_v, are given directly (Equations 3.125 and 3.126). With c_v known as a function of temperature, a simple integration gives the phonon contribution to the entropy, S. With U and S known, the Helmholtz free energy, F, follows at once. Finally, a number of thermodynamic quantities follow from F.

4

Experimental Methods

The discussion of experimental methods has been deferred until this point, because relevant background material was covered in Chapters 2 and 3. The usual experimental objectives are to determine one or both of the following quantities: 1. The elastic constants or, equivalently, the ultrasonic velocities; and 2. The dissipation or loss. The second property is known by several names, often depending on the method of measurement – ultrasonic attenuation, internal friction, logarithmic decrement, inverse Q, *etc.* Many different experimental techniques have been developed over the years, and there are various ways to categorize them. One possible approach is to divide the different methods into continuous wave techniques, in which a standing wave resonance is set up in the specimen, and pulse methods in which a short pulse of ultrasound is sent through the specimen, sometimes called the pulse-echo technique. However, when considering the problems of sample preparation and orientation, transducers, and corrections for non-ideal experimental configurations, it seems better to divide the methods into: 1. plane-wave propagation methods; and 2. methods *not* based on the plane-wave approximation.

In the discussion to follow, a few seminal papers will be cited, but the emphasis will be on modern techniques. It is not possible to cite the many, many scientists who have contributed to the development of the present-day methods.

4.1 Plane-Wave Propagation Methods

The plane-wave propagation methods are divided naturally into pulse techniques and continuous wave (resonance) techniques. After a short discussion of common problems, the pulse and resonance methods will be discussed separately below.

Figure 4.1 shows a typical sample-transducer arrangement used in the plane-wave propagation methods. Configurations with a single transducer, the same one

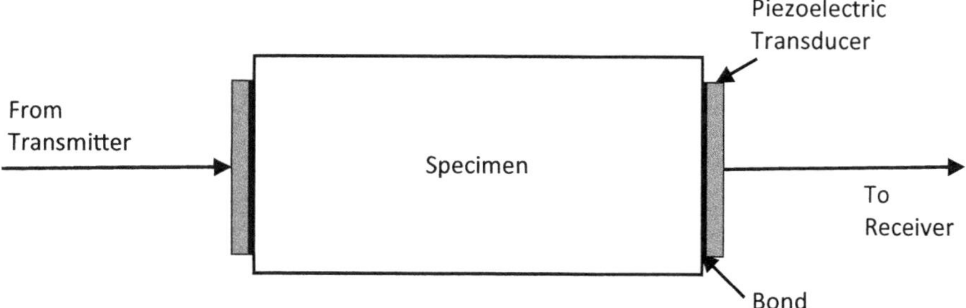

Figure 4.1 A typical transmission mode, sample-transducer assembly, with piezoelectric transducers bonded to each end of the specimen. Arrangements with only one transducer, the same being used for both excitation and reception, are common.

being used for transmitting and receiving, are also used. The specimen is prepared with flat and parallel[1] end faces. The transducers, which convert electrical voltages to mechanical displacements and *vice-versa*, are usually specially oriented cuts of piezoelectric materials and are commercially available. Single-crystal quartz, PZT, and lithium niobate are common transducer materials. Polyvinylidene fluoride (PVDF) piezoelectric film transducers have also been used [68]. Transducers are available for generating either longitudinal or transverse vibrations. Rather than piezoelectricity, transducers are sometimes based on magnetostriction or other physical effects. Many different greases, oils, and glues have been used to bond the transducer to the specimen [69], depending on experimental conditions, especially temperature. In addition to the materials listed in Ref. [69], Superglue and Crystalbond have also been used successfully. There is some art involved in producing a good bond, which should be as thin as possible. Details depend on the properties of the specimen, but the transducer is usually applied to the specimen with a twisting (wringing) motion to form the bond.

The method just described, in which a properly oriented slice of piezoelectric material is bonded to a specimen, is commonly used in the tens of MHz frequency range, but becomes difficult at higher frequencies. At UHF and microwave frequencies a better technique is to directly fabricate a thin film piezoelectric transducer onto the specimen or onto a buffer rod [70, 71, 72, 73, 74].

Several practical problems having to do with sample preparation, transducers and bonds must be addressed if reliable, accurate results are to be achieved. Fortuntely, many of these problems have been carefully discussed in earlier works [1, 2].

[1] Typically ± 0.001 mm/mm or better.

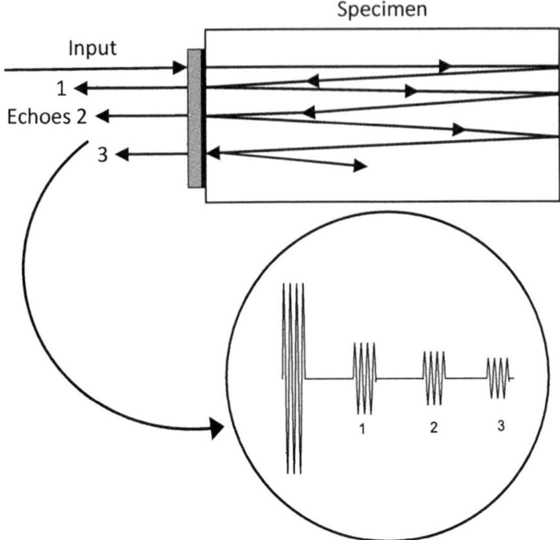

Figure 4.2 Schematic of a typical pulse-echo arrangement. The arrows drawn within the specimen are mis-oriented for display purposes. They are used to indicate the multiple reflections of the plane waves which generate the echoes. The inset illustrates the received signals showing the excitation pulse and the first three echoes.

4.1.1 Pulse-Echo Methods

A typical pulse method is illustrated in Figure 4.2. This arrangement is known as the pulse-echo method [1, 2] for obvious reasons. Other terms are sometimes used for special adaptations of the method, but the term pulse-echo will generally be used in this work. A comprehensive review of the various pulse methods has been given recently [75].

The transducer is excited by a suitable electrical signal, usually a short, high-amplitude pulse, or a short burst of a sinusoidal wave. The resulting ultrasonic pulse reverberates within the specimen, generating an echo upon each return to the transducer. Typically, most of the energy is reflected at the transducer-specimen interface and the continuing reverberation produces a string of echoes. The desired information is contained in the spacing of the echoes – roughly $\Delta t = 2l/v$ where l is the specimen length and v is the ultrasound velocity – and the decay of the echo amplitude with time, which gives the loss. Both the loss and the velocity will be discssed in more detail.

Attenuation

Loss may be introduced by assuming that the propagation vector, $\vec{K}$, of Equation 3.4 is complex. Taking z as the propagation direction and making the change $K_z \rightarrow K_z + i\alpha$ results in

$$u_z = u_z^o \exp(-\alpha z) \exp(i(K_z z - \omega t)), \tag{4.1}$$

where α is usually called the attenuation, in this case the *amplitude* attenuation.

The amplitude A of the signal is modulated by the $\exp(-\alpha z)$ term, giving

$$A(z) = A_o \exp(-\alpha z). \tag{4.2}$$

By measuring the amplitudes of two different echoes, *e.g.*, 1 and 2, α may be determined,

$$\alpha = \frac{\ln\left(\frac{A(z_1)}{A(z_2)}\right)}{(z_2 - z_1)} \tag{4.3}$$

where α is given in terms of nepers per unit length. Often the length is quoted in terms of cm. It is also common to quote the attenuation in terms of decibels in which case

$$\alpha(\text{dB/unit length}) = 20\frac{\log\left(\frac{A(z_1)}{A(z_2)}\right)}{(z_2 - z_1)}. \tag{4.4}$$

It follows that

$$\alpha(\text{dB/cm}) = 20\log(e)\alpha(\text{neper/cm}) = 8.686\alpha(\text{neper/cm}). \tag{4.5}$$

One sometimes sees $\alpha(\text{dB}/\mu\ \text{sec})$. The velocity of ultrasound is used to convert between per unit length and per unit time. In cases where several echoes are observed, it is sometimes convenient to fit Equation 4.2, an exponential decay, to the amplitudes of the string of echoes to determine α.

As will be shown elsewhere, *e.g.*, Section 4.1.2, the loss can be determined by other means such as measuring the Q of a mechanical resonance. $1/Q$ is commonly called the internal friction and is related to α in units of inverse length by $Q = \omega/(2\alpha v)$. It is a challenging task to measure the intrinsic value of α because many uninteresting factors may contribute to the attenuation of the wave. Some of these are: loss in the bond, loss to liquids or gases in which the specimen is immersed, diffraction loss, *etc.* Fortunately, the losses are additive, and one is often interested in the attenuation as a function of some variable such as the temperature, magnetic field, *etc.* To the extent that the extraneous losses may be regarded as constant,[2] it is possible to extract the loss of interest.

Velocity

A superficial analysis would suggest that the velocity of sound could be determined by measuring the time interval between successive echoes and equating that to $2l/v$

[2] Depending on the bond material and the temperature range, the *bond* loss may have a significant temperature dependence.

where l is the specimen length and v is the velocity. This oversimplified method will indeed give an approximate result for v, but will likely be in error, perhaps by a few percent. Several factors must be taken into account to achieve high accuracy. First, it is not trivial to obtain an accurate measurement of the echo to echo separation. Second, there are factors other than $2l/v$ that affect the echo to echo separation.

The shapes of the echoes (amplitude *vs.* time) change with propagation and reflection, making it difficult or impossible to find a leading edge or characteristic point on the echoes to use for an accurate measurement of the transit time [76]. Several methods have been developed to address the difficulty of measuring the transit time accurately [75]. Some of these will be discussed below. The older methods were, of necessity, based on analog electronics. It is now possible to replace the analog approach with a digital one. The digital approach offers many advantages and will be discussed in what follows. The complicating effects of phase shifts due to transducers, diffraction, bonds, and echo mismatches are basically the same in the analog and digital methods and will be discussed in a separate section.

Pulse Superposition Method

Figure 4.3 shows a simplified example of the pulse superposition method [77, 78]. In this example, only two excitation pulses are applied, but their spacing in time is equal to the time between successive echoes. The result is that echoes A2 and B1 appear at the transducer at the same time and are physically superimposed in the specimen. The same argument holds for following pairs of echoes.

In practice, rather than just two excitation pulses, the pulses are usually repeated with a time interval T, where T is approximately equal to $p\delta$. Here, p is an integer

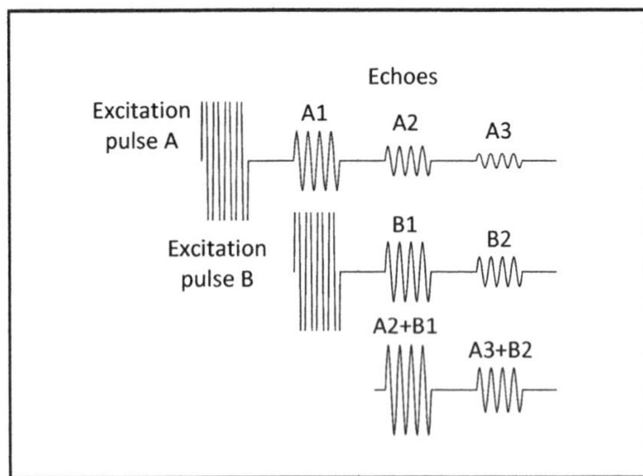

Figure 4.3 An example of the pulse superposition method for just two excitation pulses. The time between the two excitation pulses is equal to the time between successive echoes.

and δ is the desired round trip travel time for *plane waves*, unaffected by other things such as phase shifts due to the transducer or bond, diffraction, *etc.* When $p = 2$ the sum of the odd-numbered echoes appears between the excitation pulses. Other pulse sequences are also used [77].

The repetition rate, or equivalently T, is adjusted so that the sum of the superimposed echoes is a maximum. However, as described by McSkimin [77], for various reasons one cannot be sure that the observed maximum corresponds to the correct cycle-to-cycle match between the echoes. This point will be discussed later.

Pulse Echo Overlap Method

The pulse echo overlap method is similar in many ways to the pulse superposition method, but whereas there is a physical superposition of (usually) *several* echoes in the superposition method, in the pulse echo overlap method, the overlap of any *two* echoes occurs on an oscilloscope display [79]. There is no physical overlap in the specimen. A continuous wave (CW) oscillator, whose period of oscillation is approximately equal to the travel time between the two echoes of interest, serves as the x-axis drive of an oscilloscope. Suppose the echoes of interest are the first and second echo. Then the first echo appears on one sweep of the oscilloscope and the second echo on the next sweep. Of course, the third, fourth, etc. echoes will also appear on following sweeps, with the result that many echoes appear on the screen at approximately the same place. The trick is to turn down the intensity of the oscilloscope display so that no echoes are observed, and then apply appropriately time-delayed signals to the z axis to intensify the signals of the desired echoes. Then it is possible to observe only the two echoes of interest on the oscilloscope screen. The oscillator frequency is divided by a large number, ≈ 1000, before being used to trigger the pulsed oscillator to generate the ultrasonic signals; this lower repetition rate allows the string of echoes in the specimen to die out before another ultrasonic pulse is generated.

An example of the use of the pulse echo overlap method is shown in Figure 4.4. The overlap is adjusted by tuning the frequency of the CW oscillator. The inverse of the frequency for the proper overlap gives the desired transit time, T.

Digital Pulse Echo Overlap Method

A digital pulse echo method offers several advantages over the conventional methods just described [75, 80, 81]. Because the signal is recorded digitally, the time T between echoes can be determined by off-line signal processing, not by pulse superposition or echo overlap. With no need for pulse superposition or echo overlap, the electronics are simpler. Accuracy is improved because the subjective adjustment of the oscillator repetition rate to achieve the proper pulse superposition or echo

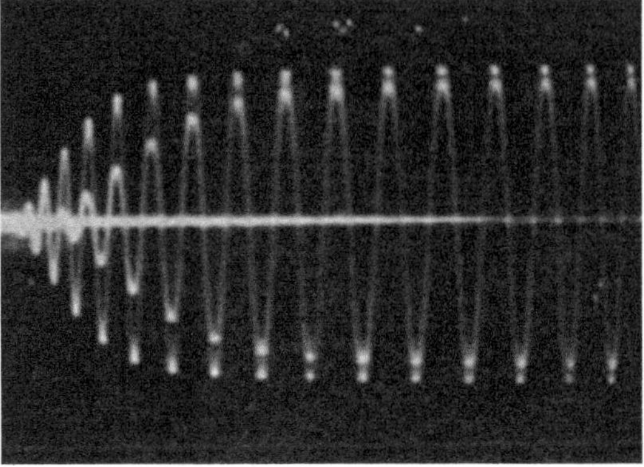

Figure 4.4 An example of the pulse-echo overlap method by Papadakis [79]. A method to determine the correct cycle-to-cycle overlap is discussed in the text.

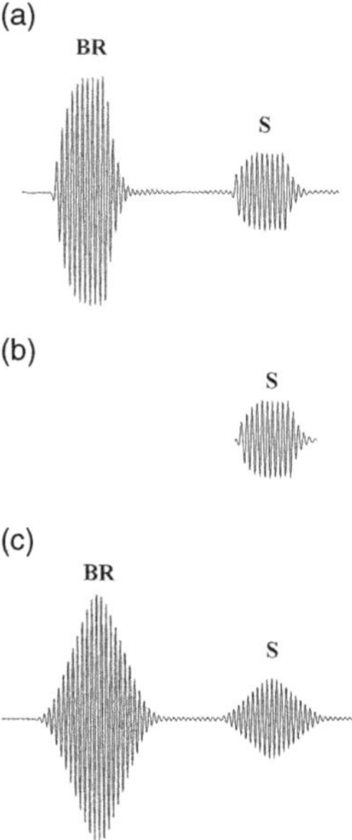

Figure 4.5 Illustration of the processing of the data from Ref. [80]. In this case a buffer rod was used. (a) BR indicates an echo from the buffer rod-specimen interface. S indicates an echo that has made one round trip through the specimen. (b) S is the same specimen echo. (c) The cross-correlation of (b) with (a).

overlap is replaced by off-line mathematical operations on the digital data. Figure 4.5, taken from Ref. [80] illustrates the use of the method. The data shown in Figure 4.5 were acquired with a buffer rod (BR) set-up. In the BR arrangement a transducer is bonded to a buffer rod of good ultrasonic properties, and the specimen is bonded to the buffer rod. Part (a) of Figure 4.5 shows one echo (BR) which has

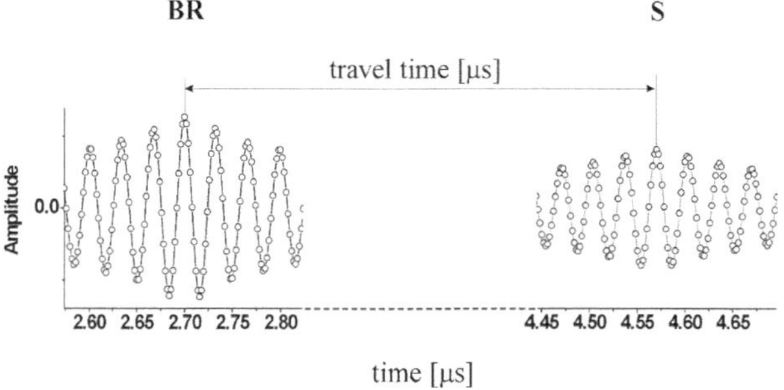

Figure 4.6 Illustration of the determination of T from the cross correlated signals of Figure 4.5, taken from [80].

traversed the buffer rod, reflected from the BR-specimen interface, and returned to the transducer. The second echo (S) has traversed the buffer rod, crossed into the specimen, traversed the specimen, reflected from the free end of the specimen and returned to the transducer. The problem is to determine the time separation between BR and S. In Ref. [80] this is achieved by using the cross correlation function, defined as

$$Corr(g,h) = \int_{-\infty}^{\infty} g(\tau + t)h(\tau)d\tau, \tag{4.6}$$

where $g(\tau)$ and $h(\tau)$ are two arbitrary functions. In the present case a section of the signal containing S duplicated and shown in part (b) of Figure 4.5. The cross correlation is performed between part (a) and part (b) of Figure 4.5 with the result shown in part (c). The cross correlation signals for both BR and S show Gaussian-like profiles.

The time delay is provisionally taken as the time between the maxima of the two cross-correlated functions. The result is shown[3] in Figure 4.6. (In cases where the transducer is bonded directly to the specimen – Figure 4.2 – the echoes may be somewhat altered in shape by reflections from the transducer, and the choice of the correct maxima in the cross correlated signals may not be obvious. This point will be addressed shortly. The present purpose is not to compare the BR arrangement with other approaches, but to demonstrate how to determine T in the digital pulse echo method.)

Relation of Phase Velocity to Measured Time between Echoes

Considering for now only the case where the transducer is bonded directly to the specimen, Figure 4.2 (as opposed to the use of a buffer rod) the various factors

[3] A fourth degree polynomial, not shown, was fitted to the central peak of BR and S for a precise determination of this time T.

affecting the time delay are considered. What follows applies to the three methods just discussed: pulse superposition; pulse echo overlap; and, digital pulse echo. The measured time interval between the selected echoes is given by [77, 2, 75]

$$T = p\delta + p\Delta t(\phi) + \frac{n}{f} - t_D. \tag{4.7}$$

As mentioned, p is an integer representing the extra number of round trips the second echo has made with respect to the first. The desired quantity is $\delta = 2l/v$, where v is the *plane-wave* sound velocity, the velocity related to the elastic constants in Section 3.1. Moving to the second term, $\Delta t(\phi)$ is a time interval due to a phase shift resulting from reflections at the sample-transducer interface, including the effect of the bond, as well as reflections at the free end of the specimen. Depending on ϕ, $\Delta t(\phi)$ can be either positive or negative. In the third term the integer n can take on either positive, negative, or zero values; the desired value is zero. This term has to do with incorrect match of the echoes in the pulse superposition or pulse-echo overlap methods; each cycle of mismatch will contribute a time error of one period of the sound wave $= 1/f$. In the digital pulse echo method, this term results from choosing an incorrect maximum in the cross correlated signal. The final term, t_D, is a correction due to diffraction. In many cases the diffraction is neglected as compared to the other terms.

If the properties of the transducer and bond are known, the term $\Delta t(\phi)$ can be calculated and taken into account [77] as shown in Appendix A, Equation A.21. If the dimensions of the transducer and specimen are known, the last term t_D, involving diffraction, may also be calculated (or perhaps neglected) as shown in Appendix B. Note that $\Delta t(\phi)$ and t_D are frequency dependent. Once these two terms are accounted for, the only remaining unknown on the right-hand side of Equation 4.7 is n. The integer n can be deduced by making measurements as a function of frequency [77, 78], being careful to maintain the same overlap condition for the chosen echoes in the pulse superposition or pulse-echo overlap methods. The digital pulse echo method offers the luxury, and power, of off-line signal processing. One can choose a particular maxima in the cross-correlated signal, as in Figure 4.6, and plot the results as a function of $1/f$, then choose a different maxima and repeat. This method offers a clear way to determine n [80]. Obviously the term involving n/f will be zero with the correct match of echoes ($n = 0$).

Buffer Rod Method

Figure 4.7 illustrates a typical pulse echo arrangement using a buffer rod [82, 73, 80]. The echoes to be compared are BR, an echo resulting from a reflection at the BR-specimen interface, and S, an echo which, as compared to BR, has made a round trip through the specimen. Figures 4.5 and 4.6 result from an arrangement

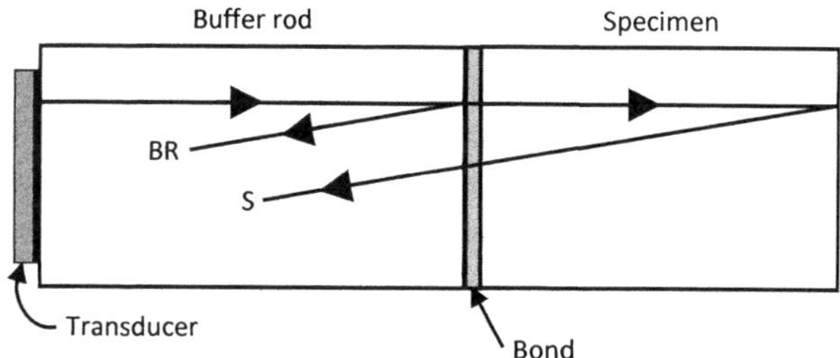

Figure 4.7 Schematic of a typical buffer rod arrangement. The echoes to be compared are BR (reflected from the buffer rod-specimen interface) and S (an echo that has traversed the specimen).

such as illustrated in Figure 4.7. Equation 4.7 applies to the present case as well, except that $\Delta t(\phi)$ has a different origin; it is related to the bond between the buffer rod and the specimen. The transducer has no effect because it affects the two echoes in the same way. If the properties of the bond are known, $\Delta t(\phi)$ can be calculated. Detailed results are given in Refs. [82, 73, 80].

Because the bond can be made very thin, $\Delta t(\phi)$ can be made much smaller than is the case when the transducer is involved. In fact, under favorable conditions, both $\Delta t(\phi)$ and t_D can be neglected initially. Then, plotting the measured time interval T *vs.* $1/f$ for various choices for the cycles of mismatch gives straight lines, yielding a clear indication of the n values for the mismatch, and enabling the correct match for the desired condition of $n = 0$ [80].

4.1.2 Plane-Wave Resonators

Ultrasonic resonators ($\approx$ 10 MHz) were used extensively to measure elastic constants of a variety of solids [83, 84, 85]. An illustration of an ultrasonic resonator is shown in Figure 4.1. Resonators are also possible with only one transducer, the same transducer being used for transmission and reception. The resonator of Figure 4.1 operates by driving one transducer with a continuous wave (CW) voltage source and detecting the response with the second transducer. At certain frequencies, resonances are observed corresponding to standing waves in the composite resonator consisting of specimen, transducer, and coupling film (bond).

The high Q of the resonators for many materials implies a great sensitivity to *changes* in the Q or equivalently the ultrasonic attenuation. In some cases the electrical impedance of the transducer-specimen assembly was arranged to control an oscillator circuit [86, 87], in which case the system was able to detect exceptionally small changes in the ultrasonic attenuation in the specimen, as small as 10^{-8} cm^{-1}, allowing the measurement of the absorption of phonons by nuclear spins. With the use of thin-film transducer technology, the CW resonator was extended to 1 GHz to detect the absorption of phonons by electrons spins [72, 88].

Plane-wave resonators are one-dimensional systems in that the only spatial variation is along the direction of propagation of the waves. An ideal resonator will be analyzed first, and then correction terms will be discussed.

Isolated Resonator

Neglecting for the moment the contributions of the transducer and bond, the resonator is easily analyzed using the concept of acoustic impedance introduced in Appendix A. Assuming the specimen of length l is terminated in $Z(l) = 0$, an excellent approximation for air, then the input impedance at the opposite end is

$$Z_{in} = Z_o \tanh(\gamma l), \tag{4.8}$$

where $\gamma = ik + \alpha$. Expanding the hyberbolic tangent gives,

$$Z_{in} = Z_o \frac{\tanh(\alpha l) + i \tan(kl)}{1 + i \tanh(\alpha l) \tan(kl)}. \tag{4.9}$$

Assuming $\alpha l << 1$, using $\tan(\alpha l) \approx \alpha l$ and rearranging terms, Equation 4.9 may be written

$$Z_{in} = Z_o \Big[\alpha l \left(1 + \tan^2(kl)\right) + i \tan(kl)\Big]. \tag{4.10}$$

The impedance is a minimum when $kl = n\pi$, where n is an integer. Using $k_n = n\pi/l = 2\pi f_n/v$ gives the resonant frequencies,

$$f_n = \frac{nv}{2l}, \tag{4.11}$$

where $n = 1, 2...$, v is the ultrasonic velocity and l is the specimen length. Equation 4.11 corresponds to $l = n\lambda/2$ where λ is the wavelength of the ultrasound. The ultrasonic velocity is given as

$$v = 2l\Delta f = \frac{2lf_n}{n} \tag{4.12}$$

with Δf being the separation between resonant frequencies.

To explore the resonant behavior it is convenient to make a change of variables, $f' = f - f_n$. Using $k = 2\pi f/v$,

$$Z_{in} = Z_o \Big[\alpha l \left(1 + \tan^2(2\pi f' l/v)\right) + i \tan(2\pi f' l/v)\Big]. \tag{4.13}$$

Assume that f' is very close to a resonance, f_n, then $2\pi f' l/v << 1$ and the tangent may be expanded so that Equation 4.13 to a good approximation may be written as

$$Z_{in} = Z_o \Big[\alpha l + i(2\pi f' l/v)\Big]. \tag{4.14}$$

In order to make an analogy with an electrical circuit it is convenient to write,

$$f' = f - f_n = \frac{(f - f_n)(f + f_n)}{(f + f_n)} \approx \frac{(f^2 - f_n^2)}{2f}. \tag{4.15}$$

Combining Equations 4.14 and 4.15 and using $Z_o = \rho v$ gives

$$Z_{in} = \rho v \alpha l + i\left[(\rho \pi l)\frac{f^2 - f_n^2}{f}\right]. \tag{4.16}$$

Equation 4.16 has exactly the form of the impedance for a series RLC circuit if the following identifications are made:[4]

$$R = \rho v \alpha l, \qquad L = \frac{\rho l}{2}, \qquad C = \frac{2l}{\pi^2 n^2 \rho v^2}, \qquad f_n = f_o. \tag{4.17}$$

Note that

$$f_o = \frac{1}{2\pi\sqrt{LC}} = \frac{nv}{2l} = f_n. \tag{4.18}$$

The quality factor for an RLC circuit is given by $Q = 2\pi f_o Ł/R$ leading to the result for the acoustic case,

$$Q = 2\pi f_n \frac{\rho l/2}{\rho v \alpha l} = \frac{2\pi f_n}{2\alpha v} = \frac{\omega_n}{2\alpha v}. \tag{4.19}$$

In summary, using the concept of acoustic impedance the ideal plane-wave resonator has been analyzed. Possible effects of transducers or coupling films on the resonances have been ignored for now. It is found that standing wave resonances are observed at frequencies given by Equation 4.11. Measurement of these frequencies and their separation yields the velocity of sound by Equation 4.12. Equation 4.19 relates the quality factor, Q, of the resonance to the amplitude attenuation coefficeient, α, of the propagating wave.

Composite Resonator

For accurate results for the phase velocities, the effects of transducers and bonds should be taken into account. The transducer corrections are treated in Appendix C. The results are expressed in terms of: $f_n^s = nv_s/2l_s$, the nth resonance of the isolated specimen; f_n^C the nth resonance of the composite resonator, the frequency actually measured; ρ_T and ρ_s, the densities of the transducer and specimen respectively; and, l_T, l_s the thicknesses of the transducer and specimen respectively.

[4] The quantities R (resistance), L (inductance), C (capacitance), as well as Z are all for unit area.

For the case of identical transducers on opposite ends of the specimen,

$$f_n^s = f_n^C + 2\left(\frac{\rho_T l_T}{\rho_s l_s}\right)(f_n^C - f^T), \tag{4.20}$$

and

$$f_{n+1}^s - f_n^s = \Delta f^s = (f_{n+1}^C - f_n^C)\left(1 + 2\left(\frac{\rho_T l_T}{\rho_s l_s}\right)\right). \tag{4.21}$$

For the case of a single transducer attached to one end of the specimen

$$f_n^s = f_n^C + \left(\frac{\rho_T l_T}{\rho_s l_s}\right)(f_n^C - f^T), \tag{4.22}$$

and

$$f_{n+1}^s - f_n^s = \Delta f^s = (f_{n+1}^C - f_n^C)\left(1 + \left(\frac{\rho_T l_T}{\rho_s l_s}\right)\right). \tag{4.23}$$

The effect of a bond is not treated here, but is usually much less than that of the transducer [83], at least in the 10 Mhz range. At very high frequencies the situation may be different.

The output of a plane-wave resonator, using thin-film transducers directly deposited onto the specimen, is shown in Figure 4.8. Ten resonances $f_n^C \approx f_n^s$ are shown in the top part of the figure. This system was used to detect the absorption of phonons by electron spins, and operated down to 4 K [88]. The overall envelope in the top part of Figure 4.8 is due to the impedance matching, not to the transducer bandwidth.

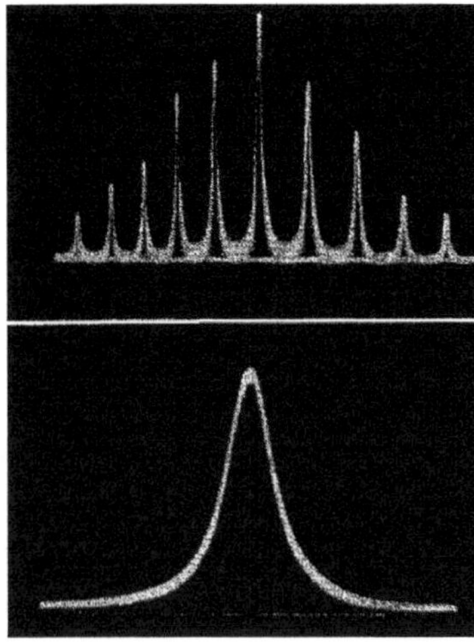

Figure 4.8 Resonances of 1.1 GHz longitudinal waves propagating along a cube axis in MgO at 77K. Top. Sweep width approximately 7.0 MHz. Bottom. Sweep width approximately 0.6 MHz. From Ref. [72].

4.1.3 Limitations of the Plane-Wave Approximation Approach

Both the pulse-echo and resonance methods just discussed allow quick, easily interpreted results. However, these methods have certain limitations.

- Plane wave solutions for various symmetries were discussed in Section 3.1 where relations between ultrasonic velocities and elastic constants were given. These relations are fairly simple for cubic symmetry, but become increasingly complicated for lower symmetries. Thus for cubic symmetry it is possible to obtain all three elastic constants by propagation of one longitudinal and two transverse waves along the [110] direction; but, orthorhombic symmetry requires nine different velocities with measurements along six different directions. Many interesting materials have rather low symmetry and it may be impractical to determine the full set of elastic constants by the methods just discussed.
- To limit diffraction effects the diameter of the transducer should be much greater than the wavelength of sound in the specimen. There will still be some diffraction, so to limit effects due to the ultrasonic beam scattering from the sides of the specimen, the specimen diameter should be somewhat greater than the transducer diameter. In addition, in order to resolve individual echoes in the pulse-echo method, a minimum sample thickness is required. For frequencies of 10 MHz, these considerations lead to specimen diameters of the order of one cm. Single crystals of the most interesting materials are often unavailable in such large sizes.
- The effects of the transducer and bonds on the measured time intervals, or on the resonant frequencies, must be accounted for if accurate results are to be obtained. In addition the bond between specimen and transducer often produces unwanted effects such as coupling losses and non-parallelism of the ends. The usual transducer is phase sensitive; the voltage out is proportional to the sum of the amplitudes, not the intensities, of the waves impinging across the face of the transducer. A poor bond may produce cancellation effects which distort the true signal. Temperature dependence of the bonds is often a practical problem. This problem can sometimes be solved by using the same bond material on a known material over the temperture range of interest.

4.2 Resonant Ultrasound Spectroscopy

In contrast to the plane wave propagation methods, there are other methods for measuring elastic properties that differ fundamentally from those just described in that they do not depend on plane wave propagation at all. These methods include the resonant excitation of torsional, longitudinal, and flexural vibrations of long thin bars [29, 89, 90, 91]. Measurement of the resonant frequency gives the shear modulus for the torsional case, and Young's modulus for the next two cases.

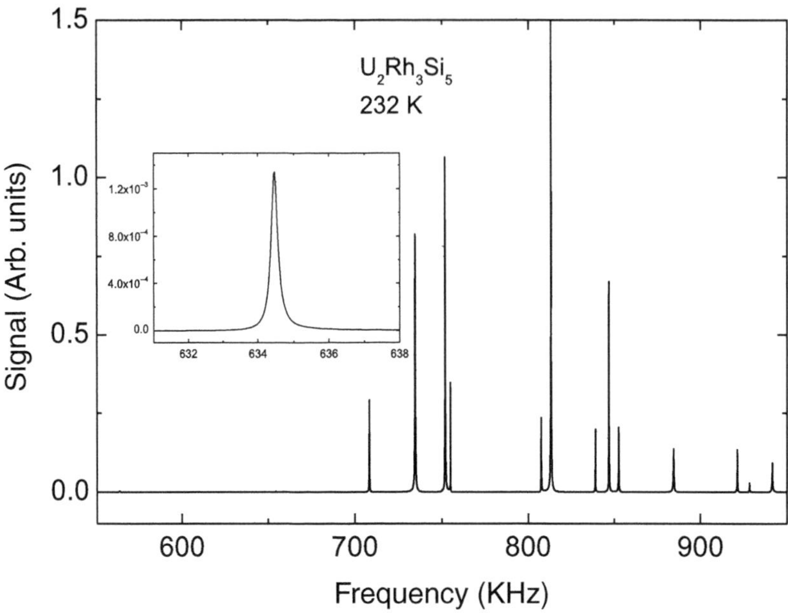

Figure 4.9 A typical RUS spectrum, in this case for a crystal of $U_2Rh_3Si_5$. The inset shows a resonance that is too small to be visible in the main figure.

A great advance in this area was the development of the technique known as resonant ultrasound spectroscopy (RUS) [92, 93, 94]. RUS is based on the measurement of the vibrational eigenmodes of samples of well-defined shapes. The method is not restricted to special shapes. A sample, often a parallelepiped, is held lightly between two piezoelectric transducers. In contrast to the case of Figure 4.1, the transducer is *not* bonded to the specimen, but touches the specimen lightly, ideally at a point.[5] The sample is excited at one point by one of the transducers. The frequency of this driving transducer is swept through a range corresponding to a large number of vibrational eigenmodes of the sample. The resonant response of the sample is detected by the opposite transducer. A segment of a typical RUS spectrum is shown in Figure 4.9. The spectrum is very different from that of the simple 1D resonator of Figure 4.8; in particular, the RUS resonances are not equally spaced in frequency.

As will be discussed below, the eigenfrequencies depend on the elastic constants, the sample shape, the orientation of the crystallographic axis with respect to the sample, and the density. By measuring a large number of resonant frequencies on one sample it is possible to obtain information about all these quantities.

[5] Earlier work showed the specimen mounted corner-to-corner between the transducers, but this difficult balancing act has been found to be unnecessary, and even has some deleterious effects [95]. Even with a "flat" contact and no bonds, imperfections in transducers and specimens usually lead to point excitation.

(The frequencies scale with the square root of elastic constants and the linear dimensions of the sample; thus, elastic constants and linear dimensions cannot be determined independently.) Usually the sample shape, crystallographic orientation, and density are known and one determines the complete elastic constant matrix from such a spectrum.

For accurate results careful attention must be paid to a number of factors, including sample preparation, electronics, and transducer design. Fortunately, these issues have been addressed in [94] and [95].

It is apparent that RUS overcomes many of the deficiencies of the more conventional methods for the measurement of elastic constants [96].

- There are no diffraction effects to worry about; there is no plane-wave approximation. The inherent accuracy is high.
- Because there is no need to establish plane-wave propagation, small samples are naturally accommodated. Sample dimensions of approximately 1 mm typically result in many eigenfrequencies below 4 MHz, a convenient range. Accurate measurements on sub-mm specimens are feasible. Compared to more conventional methods described above, the minimum sample volume is reduced by roughly three orders of magnitude. Many novel materials, especially in single-crystal form, are only available as small samples, thus this reduction in minimum size vastly expands the range of materials susceptible to precise ultrasonic measurements.
- There is no bond between the transducer and the sample, only contact force is used. The absence of a bond is very useful for temperature dependent measurements because differential thermal contraction often leads to bonds breaking, or at best strains being applied to the sample under investigation. This absence of a bond is especially important near phase transitions.
- With sensitive electronics, a very low contact force is possible so that transducer loading effects are negligible.
- All of the elastic constants are determined from one spectrum. There is no need to prepare separate samples with different crystalline orientations. As a result, high relative accuracy for the elastic constants is obtained. Low-symmetry materials are accommodated almost as easily as high-symmetry materials.

Figure 4.10 shows several computed eigenmodes of a parallelepiped specimen. This Figure helps explain the very wide range of amplitudes of the resonances shown in Figure 4.9. The usual situation is that compressional mode transducers are used for excitation and detection. Thus, only the component of specimen motion *perpendicular* to the surface of the transducer is effective in producing and detecting vibrations. Figure 4.10 shows that the mode type and the point at which the transducer contacts the specimen each have a strong effect on signal amplitude. The

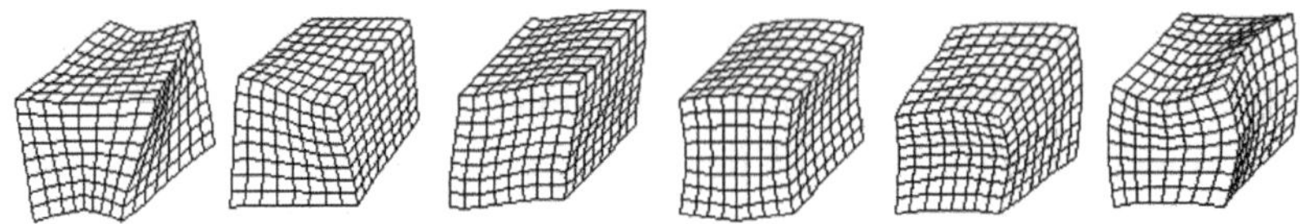

Figure 4.10 Some typical RUS vibrational eigenmodes for a parallelepiped. The vibrational amplitudes are, of course, highly exaggerated.

inset in Figure 4.9 shows a resonance at 634.5 kHz that is not visible on the main scan; the ratio of the amplitude of the highest signal to the one shown in the inset is ≈ 1000. The signal at 634.5 kHz is almost "silent," presumably because the motion of the contact point is almost parallel to the surface of the transducer or the point of contact happens to be at a vibrational node. In cases where an expected resonance is not found, it is good practice to simply remove and replace the specimen between the transducers as it is unlikely that the same contact point and orientation will be reproduced, and the expected signal may be found.

It is probably no accident that RUS has achieved prominence at a time when the computation power of personal computers has increased dramatically. As suggested by Figure 4.10, the vibrational eigenmodes of a three-dimensional object are rather complicated. A RUS spectrum contains much information, but extracting that information is not a simple task. The procedure is to compare the measured frequencies with computed frequencies, which are calculated with an initial set of input parameters. The input parameters for the computation are varied in an iterative process to obtain good agreement between measured and computed frequencies, and in this way the input parameters are determined. Usually the input parameters are the elastic constants, although parameters describing crystalline orientation or sample shape can be used. Thus, an essential ingredient of RUS is the ability to calculate the eigenfrequencies in an efficient manner and systematically iterate toward a match of the measured and computed frequencies.

4.2.1 Calculation of the Vibrational Eigenmodes

An integral element of RUS is comparison of measured and computed frequencies. The direct approach would be to solve the equation of motion, Equation 2.29, for the eigenfrequencies; however, an exact solution for a 3D object remains unsolved, except for some highly special cases. An approximate solution is needed. Equation 2.29 is not the most convenient starting point for such an approximation. The methods adopted will be discussed below, after a brief summary regarding Lagrange's equations for an elastic continuum.

Starting with early work on cube resonances [97, 98, 99], the computational methods have been developed to include parallelepipeds [92], cylinders [100], and arbitrary shapes [101, 93, 94]. The method is based on Hamilton's principle which states that the motion of a system over a time interval from t_1 to t_2 is such that the time integral of the Lagrangian is stationary. That means that for a small change in the path followed by the system from t_1 to t_2

$$\delta \int_{t_1}^{t_2} L dt = 0, \tag{4.24}$$

where $L = T - V$ is the Lagrangian and T and V are the kinetic and potential energies respectively. The general subject of a stationary path of an arbitrary integrand f between two points is treated in the calculus of variations. The conditions for a stationary path are known as Euler's equations [9]. Thus, the Lagrange equations are just a special case of the Euler equations for the case $f = L$. The usual situation in standard classical mechanics is $L = L(x_1, x_2, \ldots., x_n, \dot{x}_1, \dot{x}_2, \ldots.., \dot{x}_n, t)$, where $\dot{x}_i = dx_i/dt$. The resulting Lagrange equations are

$$\frac{d}{dt}\frac{\partial L}{\partial \dot{x}_i} - \frac{\partial L}{\partial x_i} = 0, \tag{4.25}$$

for each x_i. Thus, Equations 4.25 are the conditions for L to be stationary.

The situation is more involved for continuous media. The first, simple, difference is that we deal with the Lagrangian density $\mathcal{L}$, where

$$L = \int \mathcal{L} dV \tag{4.26}$$

Consider now a single component of the displacement u_i. In the problem of waves in a continuous medium we have

$$\mathcal{L} = \mathcal{L}\left(u_i, \frac{\partial u_i}{\partial x_1}, \frac{\partial u_i}{\partial x_2}, \frac{\partial u_i}{\partial x_3}, \frac{\partial u_i}{\partial t}, x_1, x_2, x_3, t\right). \tag{4.27}$$

In addition to the time derivative there is now a spatial derivative. Thus, x and t appear on an equal footing in the Lagrangian density, and must appear on an equal footing in the Lagrange equations. In this case the Lagrange equations are [9, 102],

$$\frac{d}{dt}\frac{\partial \mathcal{L}}{\partial \dot{u}_i} + \sum_j \frac{d}{dx_j}\left(\frac{\partial \mathcal{L}}{\partial \left(\frac{\partial u_i}{\partial x_j}\right)}\right) - \frac{\partial \mathcal{L}}{\partial u_i} = 0, \tag{4.28}$$

for each $i = 1, 2, 3$. (To avoid overly complex notation only one component, u_i, was listed in Equation 4.27, but the result holds for all three components of u.)

Now to return to the calculation of the vibrational eigenmodes. The kinetic energy density is just $\frac{1}{2}\sum_i \rho \dot{u}_i^2$. The potential energy density is found by using the four-index form of Equation 2.69 and the symmetry of the strain tensor, Equation 2.9, in the summation,[6]

$$L_d = \int \mathcal{L}_d dV = \int \left(\frac{1}{2} \sum_i \rho \dot{u}_i^2 - \frac{1}{2} \sum_{ijkl} c_{ijkl} \frac{\partial u_i}{\partial x_j} \frac{\partial u_k}{\partial x_l} \right) dV. \tag{4.29}$$

For steady state conditions, it has been shown [103] that L_d is equivalent to

$$L_s = \int \mathcal{L}_s dV = \int \left(\frac{1}{2} \sum_i \rho \omega^2 u_i^2 - \frac{1}{2} \sum_{ijkl} c_{ijkl} \frac{\partial u_i}{\partial x_j} \frac{\partial u_k}{\partial x_l} \right) dV. \tag{4.30}$$

To show that the two expressions are equivalent for the steady state, Equation 4.28 will be applied to each of these expressions. The result for $\mathcal{L}_d$ is

$$\rho \ddot{u}_i - \sum_{jkl} c_{ijkl} \frac{\partial^2 u_k}{\partial x_j \partial x_l} = 0. \tag{4.31}$$

Equation 4.31 was found previously using Newton's Laws, Equation 3.3. Assuming that u_i varies as *e.g.* $\cos(\omega t + \theta)$ then $\ddot{u}_i = -\omega^2 u_i$ and Equation 4.31 becomes

$$\rho \omega^2 u_i + \sum_{jkl} c_{ijkl} \frac{\partial^2 u_k}{\partial x_j \partial x_l} = 0, \tag{4.32}$$

the wave equation.

Applying Equation 4.28 to $\mathcal{L}_s$ gives immediately Equation 4.32. (Equation 4.32 also follows immediately from Equation 3.3.) This exercise proves that $\mathcal{L}_d$ and $\mathcal{L}_s$ are equivalent for computing steady state results.

A formal investigation of the stationary value of L_s in a volume enclosed by a surface finds Equation 4.32 within the volume, and

$$\sum_{jkl} c_{ijkl} \frac{\partial u_k}{\partial x_l} l_j = 0 \tag{4.33}$$

on the surface [101, 68]. Equation 2.37 for the stress on a surface may be written as

$$f_i = \sum_j \sigma_{ij} l_j = \sum_{jkl} c_{ijkl} \frac{\partial u_k}{\partial x_l} l_j. \tag{4.34}$$

In both Equations l_j is the jth component of the unit vector normal to the surface. The conclusion is that the u_i giving a stationary Lagrangian correspond to a stress-free boundary.

[6] The indices i, j, k, l each range over 1,2,3.

The steady state Lagrangian L_s has been found to be the useful one for use with Hamilton's principle. The computation of the eigenmodes involves finding displacements u_i that approximate a stationary value of L_s. The procedure starts with the expansion of the displacements in terms of a complete set of basis functions,

$$u_i(r) = \sum_{\alpha} a_{i\alpha} \Phi_\alpha(r) \tag{4.35}$$

where $a_{i\alpha}$ are the expansion coefficients and $\Phi_\alpha(r)$ are the basis functions. The $\Phi_\alpha(r)$ are often chosen based on the shape of the sample. In particular, it is important to be able to integrate the basis functions and their derivatives over the volume of the sample. For parallelepipeds, a convenient choice for $\Phi_\alpha(r)$ is the Legendre polynomials [99, 92, 96], while a series expansion in powers of the Cartesian coordinates x_1, x_2 and x_3, *i.e.* $x_1^l x_2^m x_3^n$ where l, m, and n are integers, is useful for a variety of shapes [101]. Each of the three components of the displacement is expanded as in Equation 4.35 and each has its own set of expansion coefficients. Substituting Equation 4.35 into L_s yields,

$$L_s = \frac{1}{2}\left(\sum_{ik\alpha\beta} a_{i\alpha} a_{k\beta} \rho\omega^2 \int_V \delta_{ik} \Phi_\alpha(r)\Phi_\beta(r) dV - \sum_{ijkl\alpha\beta} a_{i\alpha} a_{k\beta} \int_V c_{ijkl} \frac{\partial \Phi_\alpha}{\partial x_j} \frac{\partial \Phi_\beta}{\partial x_l} dV \right). \tag{4.36}$$

Equation 4.36 can be written more compactly as

$$L_s = \frac{1}{2}\left(\rho\omega^2 a^T \boldsymbol{E} a - a^T \boldsymbol{\Gamma} a\right) \tag{4.37}$$

where the expansion coefficients $a_{i\alpha}$ are elements of a column vector, and $\mathbf{E}$ and $\boldsymbol{\Gamma}$ are matrices with elements

$$\boldsymbol{E}_{\alpha i\beta k} = \delta_{ik} \int_V \Phi_\alpha \Phi_\beta dV \tag{4.38}$$

and

$$\boldsymbol{\Gamma}_{\alpha i\beta k} = \sum_{jl} \int_V c_{ijkl} \frac{\partial \Phi_\alpha}{\partial x_j} \frac{\partial \Phi_\beta}{\partial x_l} dV. \tag{4.39}$$

The condition that L_s be an extremum is found by setting the derivatives of L_s with respect to each of the expansion coefficients equal to zero. This process yields the eigenvalue equation

$$\boldsymbol{\Gamma a} = \rho\omega^2 \boldsymbol{E a}. \tag{4.40}$$

Finding the resonant frequencies of a 3D object first involves calculating $\boldsymbol{E}$ and $\boldsymbol{\Gamma}$ using the elastic constants and the sample shape for the specimen of interest. Next Equation 4.40 can be solved by standard techniques. The eigenvalues are $\lambda = \rho\omega^2$ and the eigenvectors $\boldsymbol{a}$ are the expansion coefficients $a_{i\alpha}$.

4.2.2 Elastic Constant Determination

The first step in the determination of the elastic constants is to measure a number of resonant frequencies from a scan such as shown in Figure 4.9. Reference [95] has a good discussion of the number of frequencies needed for different situations. For sharp resonances and strong signals it is easy to determine the resonant frequencies from a display on a computer monitor, but in other cases, and for improved accuracy in any case, it is a good idea to fit the resonances to theoretical curves as described in Section 4.2.3. Next, an initial estimate is made for the elastic constants and Equation 4.40 is solved for the resonant frequencies. These computed frequencies are compared to the measured ones and an error function F is computed:

$$F = \sum_{i=1}^{N} w_i (f_i - g_i)^2, \tag{4.41}$$

where f_i and g_i are the computed and measured frequencies respectively, and w_i is a weighting factor usually taken as unity. An essential feature of the whole process is the minimization of the error function. The Levenberg-Marquardt [104] algorithm is used for this purpose [93, 96]. Fortunately, the entire RUS code is available at [105]. Discussions of practical considerations for the successful implementation of the procedure are available in several papers including [95].

4.2.3 Precision Determination of Q and f_n

In principal, the measurement of internal friction is quite simple. One simply finds the half-power points (or the points where the *amplitude* of the signal drops to $1/\sqrt{2}$ of the maximum value) for a particular resonance. Then, the internal friction is defined as $1/Q$ where $Q = \omega_n/\Delta\omega_n$ and $\Delta\omega_n$ is the separation of the half power points for resonant frequency ω_n. In practice, such an approach will not usually give high-quality results for two reasons. One, it is difficult to determine the half power points accurately. Second, in some cases there may be interfering signals that give an unwanted background to the RUS spectrum. Because the interfering signal is often coherent with the real RUS signal, it can produce a significant distortion of the latter. This situation is, of course, more problematic for highly attenuating samples, especially if the sample is small or is located in a cryostat necessitating long cables.

This interfering signal can be due to direct electromagnetic coupling between the transmitting and receiving sides of the spectrometer, sound transmission through the air or the sample-transducer assembly, or other reasons. In such cases it is best to fit the signal of interest and the background signal to a theoretical expression so as to obtain the Q *and* the resonant frequency.

A damped, driven oscillator [94, 106] will be used to model the RUS signal near a resonant frequency. The driving force is taken as $Fe^{-i(2\pi ft+\theta)}$, where $f = \omega/2\pi$ is the driving frequency[7], t is the time, and θ represents an often unknown phase shift (due to electronics, transducers, nearby resonances, *etc.*) between the drive signal and the detected signal. The equation of motion for a simple damped, driven oscillator is

$$Fe^{-i(2\pi ft+\theta)} - kx - b\frac{dx}{dt} = m\frac{d^2x}{dt^2}, \tag{4.42}$$

where kx is the usual Hooke's law force and $b\frac{dx}{dt}$ represents a velocity-dependent damping force.[8] Useful changes in notation appropriate to the present situation are: $k/m = (2\pi f_n)^2$ where f_n is the *nth* resonant frequency of the undamped, undriven oscillator; and $b/m = 2\pi f_n/Q$, where Q has the usual meaning. The *steady state* solution can be written

$$x(t,f) = L(f)e^{-i2\pi ft}. \tag{4.43}$$

The result for $L(f)$ is

$$L(f) = A\frac{e^{-i\theta}}{(f_n^2 - f^2) - i(f_n f/Q)} \tag{4.44}$$

where $L(f)$ is a Lorentzian response function in complex notation and $A = F/(4\pi^2 m)$, an amplitude not of much interest for the present problem.

The background signal is represented in terms of an expansion in frequency [107],

$$g(t,f) = B(f)e^{-i2\pi ft} \tag{4.45}$$

with

$$B(f) = b_1 + ib_2 + \frac{(c_1 + ic_2)(f - f_n)}{f_n} \tag{4.46}$$

where b_1, b_2, c_1, and c_2 are parameters describing the background. It is usually not necessary to keep terms higher than linear in the expansion of the background signal.

[7] Because the results will be used to fit experimental data, f, in Hz, will be used in this section rather than the angular frequency ω, in radians/sec.

[8] Appendix D offers an alternate way to treat the damping.

The sum of L and B is written as

$$w(f) = u(f) - iv(f). \tag{4.47}$$

It is now time to dispense with the complex notation. The detected signal is the real part of $w(f)e^{-i2\pi ft}$, which is

$$s(f,t) = u(f)\cos(2\pi ft) - v(f)\sin(2\pi ft), \tag{4.48}$$

where $u(f)$ and $v(f)$ are called the in-phase and quadrature signals, respectively. These quantities are given by

$$u(f) = \left\{ a\left[\frac{f_n^2}{Q}\right]\frac{(f_n^2 - f^2)\cos\theta + (f_n f/Q)\sin\theta}{(f_n^2 - f^2)^2 + (f_n f/Q)^2} + b_1 + \frac{c_1(f - f_n)}{f_n} \right\} \tag{4.49}$$

and

$$v(f) = -\left\{ a\left[\frac{f_n^2}{Q}\right]\frac{(f_n f/Q)\cos\theta - (f_n^2 - f^2)\sin\theta}{(f_n^2 - f^2)^2 + (f_n f/Q)^2} + b_2 + \frac{c_2(f - f_n)}{f_n} \right\} \tag{4.50}$$

where

$$a = A\left[\frac{Q}{f_n^2}\right]. \tag{4.51}$$

In the present treatment $u(f)$ is the real part of the complex response function $w(f)$ and $v(f)$ is *minus* the imaginary part of $w(f)$.

It is more convenient to use a rather than A in finding the best fit parameters. The basic Lorentzian function increases in amplitude as Q increases. This increase is offset by the factor(f_n^2/Q) in Equations 4.49 and 4.50. Thus, a and Q can be varied somewhat independently in the fitting procedure whereas the correct value of A depends inversely on Q.

The two Equations 4.49 and 4.50 are fit simultaneously to the in-phase and quadrature RUS signals for a particular resonance. The fitting parameters are $f_n, Q, a, \theta, b_1, b_2, c_1$, and c_2 of which only f_n and Q are usually of interest. The values of f_n and Q determined by fitting the in-phase signal should be very close to those found from fitting the quadrature signal. Figure 4.11 gives an example of the fitting process for a steel specimen. For this case the two values of f_n are the same, $f_n = 527.346$ kHz, and the two Q values differ by 0.3%, the average being 13,180. Note that this fitting procedure gives f_n, the frequency at which the undriven, undamped oscillator vibrates.

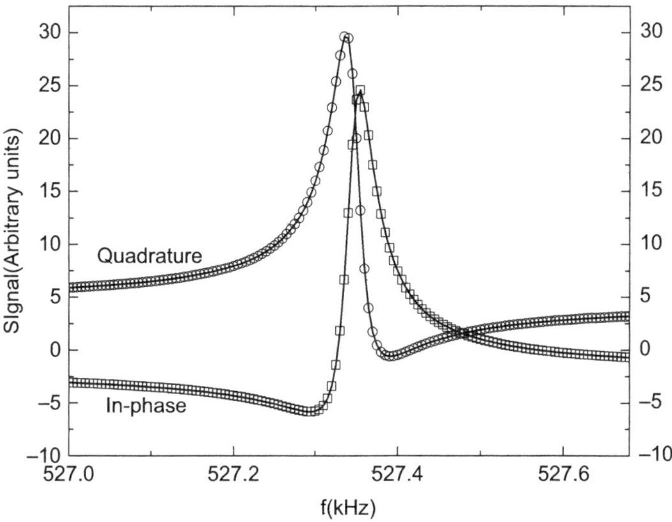

Figure 4.11 The in-phase and quadrature RUS signals for a steel specimen. The solid lines are the recorded data. The open symbols are the results of fitting Equations 4.49 and 4.50 to the data, yielding the parameters Q and f_n.

If the experiment is not phase sensitive, but simply measures the amplitude of the signal $\mathcal{A} = \sqrt{u(f)^2 + v(f)^2}$, then neglecting the background terms,

$$\mathcal{A} \simeq \frac{1}{\sqrt{(f_n^2 - f^2)^2 + \left(\frac{f_n f}{Q}\right)^2}}. \tag{4.52}$$

$\mathcal{A}$ has its maximum value when

$$f = f_{max} = f_n \sqrt{1 - \frac{1}{2Q^2}} \tag{4.53}$$

which, for large Q, differs only slightly from f_n. Presumably, f_n is the appropriate experimental resonant frequency to use in determining the elastic constants, because the computations of Section 4.2.1 are for free vibrations and no damping. For reasonably large Q there is a negligible difference between the different measures of the "resonant" frequency. Chapter 1 of [94] has an excellent description of the various "resonant" frequencies.

Neglecting terms of order Q^2, valid for moderately high Q, it is easy to show [94] that the frequencies where the amplitude $\mathcal{A}$ falls to $1\sqrt{2}$ of its maximum value (power falls to $1/2$ of its maximum value) are given by $f_\pm = f_n \pm f_n/(2Q)$. Defining $\Delta f = f_+ - f_-$ leads to the well-known result,

$$Q = \frac{f_n}{\Delta f}. \tag{4.54}$$

Using the curve-fitting procedures described in this section leads to improved precision in the determination of the f_n. Perhaps more important, this fitting gives greatly improved results for the internal friction $1/Q$, which permits sensitive internal friction measurements as a function of external parameters such as temperature [108].

4.3 Picosecond Ultrasonics

The usual pulse-echo technique requires a minimum specimen thickness so that individual echoes may be resolved, *i.e.*, the round-trip travel time should be greater than the ultrasonic pulse width. Thus, for thinner specimens, shorter pulse widths are needed. However, ultrashort pulse widths require ultrahigh frequencies. Because the ultrasonic attenuation usually increases with frequency and with temperature, it is extremely difficult to increase the ultrasonic frequency beyond a few GHz at room temperature, indeed, usual ultrasonic measurements are carried at frequencies in the 10's to 100's MHz range.

Laser generation and detection offer a path to the ultrashort pulses needed to investigate many interesting problems, *e.g.*, thin films [109, 110, 111]. With this method, a stress pulse is generated by focusing a short laser pulse ($\approx 100 fs$), the pump, on the surface of a light absorbing material. The temperature rises almost instantly within the volume determined by the area A of the laser spot and ξ, the absorption length.[9] The result is an inhomogeneous elastic stress dependent on the thermal expansion and the bulk modulus. (The stress immediately after the light is absorbed is just that required to counteract thermal expansion.) It is assumed that the diameter of the spot is $>> \xi$. With this assumption the problem is essentially one dimensional. This inhomogeneous stress profile, very near the surface, begins to relax. It can relax by moving away from the surface and also toward the surface. However, any propagation toward the surface will be reflected at the surface with a change in sign, Equation A.14. The result is an acoustic pulse propagating away at the sound velocity v. If, as assumed, the diameter of the spot is $>> \xi$ then the result is an approximately longitudinal wave. The width of the acoustic pulse is not determined by the width of the very short laser pump pulse; instead, the spatial width is approximately ξ. The pulse has a range of frequency components peaked at $\omega \approx v/\xi$, giving a pulse width in time of roughly ξ/v. Typical values for metals result in pulse widths of a few picoseconds.

Detection utilizes a low-energy laser probe pulse as illustrated in Figure 4.12. The probe pulse is operated with a variable delay. A returning echo, with its elastic

[9] In a metal or semiconductor the light may first be absorbed by electrons and carried some distance before the energy is transferred to the phonon thermal system.

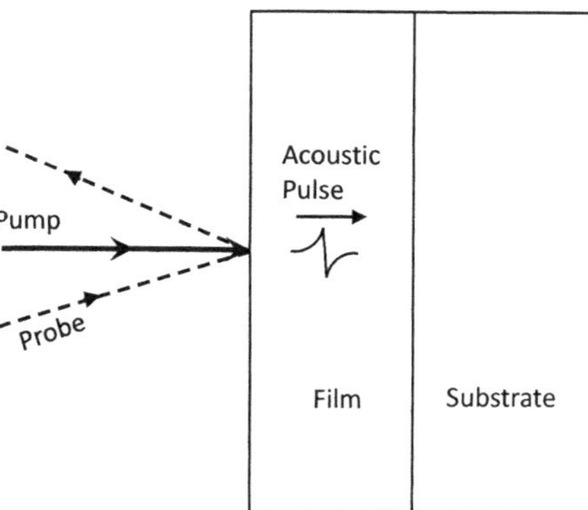

Figure 4.12 Illustration of the experimental arrangement for the laser-based generation and detection of acoustic waves. A pump laser pulse generates an acoustic pulse that travels across the thin film and partially reflects at the film-substrate interface. The returning echo alters the optical properties of the film surface and is detected by a variable-delayed laser probe pusle [111].

strain, alters the optical properties at the surface, and hence the optical reflectivity [110, 112]. Returning pulses have also been detected by changes in the surface slope, which results in a deflection of the reflected probe pulse [113]. Thus, the laser generation and detection method permits the measurement of acoustic velocities and attenuation in submicron thin films in the 100's of GHz range [114].

In cases where the medium is transparent, or partially transparent, the probe pulse penetrates deeper, and the probe pulse and the acoustic pulse propagate together. The acoustic strain produces variations in the local optical properties. As a result, part of the probe pulse is reflected by the acoustic strain, and part at the remainder at interface. These two reflections interfere constructively or destructively depending on the time-dependent strain [115]. The result is a periodic variation in the reflectivity known as Brillouin oscillations [116, 117, 118]. The method of Brillouin oscillations has been used to measure critical parameters of thin films such as acoustic attenuation and velocity, index of refraction and film thickness [114, 119, 118, 120], as well as the temperature dependence of the adiabatic sound velocity in thin films [121]. These picosecond laser methods have also been applied to the study of biological materials [122, 123].

With advancements in the laser method of generation and detection of extremely high frequency phonons, including the amplification of coherent sub-terahertz phonons [124], this technique seems destined to become an important tool for the studies of microelectronic devices, nanostructures, and biological materials as well as topics not yet imagined.

5

Elastic Constants

5.1 Introduction

The elastic constants are of fundamental importance to the understanding of materials [125]. Because they are the second derivatives of thermodynamic potentials with respect to strain, elastic constants are closely related to interatomic potentials. They are also affected by various excitations in the material. The present chapter is concerned with the physics underlying the magnitude of the elastic constants, and more especially with the variation of the elastic constants with externally controlled parameters, such as temperature. After a review of relevant background material, several applications of the central ideas will be discussed in some detail to illustrate the basic physics, but no attempt will be made to include a comprehensive discussion of the myriad phenomena affecting the elastic constants of solids.

5.2 Relevant Thermodynamics and Statistical Mechanics

Drawing on the ideas presented in Section 2.3.3, the first law of thermodynamics can be written [4]

$$dU = TdS + \sum_{m=1}^{6} \sigma_m d\epsilon_m, \tag{5.1}$$

when the work is in the form of infinitesimal elastic deformations. This form shows that the independent variables for the internal energy U are the entropy, S, and the six strains ϵ_m.

It proves useful to introduce the Helmholtz free energy defined as $F = U - TS$. It follows that

$$dF = -SdT + \sum_{m} \sigma_m d\epsilon_m \tag{5.2}$$

From the properties of a differential of a function, *e.g.*, $f(x_1, x_2, x_3, \ldots)$

$$df = \frac{\partial f}{\partial x_1} dx_1 + \frac{\partial f}{\partial x_2} dx_2 + \frac{\partial f}{\partial x_3} dx_3 + \cdots \tag{5.3}$$

it is possible to immediately obtain from Equation 5.2 the following relations

$$S = -\left(\frac{\partial F}{\partial T}\right)_\epsilon \qquad \sigma_m = \left(\frac{\partial F}{\partial \epsilon_m}\right)_T. \tag{5.4}$$

Of course, when taking derivatives with respect to a particular variable, all the other variables are held constant. It is cumbersome to fully indicate this in the above equation, particularly with the six ϵ_m, so it is simply necessary to keep this in mind. The independent variables for F are the temperature T and the six strains ϵ_m. Because F is a perfect differential the order of differentiation of F with respect to either T or ϵ_m does not matter, which results in

$$\left(\frac{\partial S}{\partial \epsilon_m}\right)_T = -\left(\frac{\partial \sigma_m}{\partial T}\right)_\epsilon = f_m, \tag{5.5}$$

where f_m is defined by Equation 5.5 [4].

It is also useful to define the Gibb's free energy, $G = F - \sum \sigma_m \epsilon_m$, from which it follows that

$$dG = -SdT - \sum \epsilon_m d\sigma_m. \tag{5.6}$$

Following the same arguments used for F it is found that

$$S = -\left(\frac{\partial G}{\partial T}\right)_\sigma \qquad \epsilon_m = -\left(\frac{\partial G}{\partial \sigma_m}\right)_T \tag{5.7}$$

and

$$\left(\frac{\partial S}{\partial \sigma_m}\right)_T = \left(\frac{\partial \epsilon_m}{\partial T}\right)_\sigma = \alpha_m \tag{5.8}$$

where the α_m are elements of the thermal expansion tensor [4]. The independent variables for G are T and the six σ_m.

From Equations 2.62 and 2.63, the following useful relations are,

$$c^S_{mn} = \left(\frac{\partial \sigma_m}{\partial \epsilon_n}\right)_S \tag{5.9}$$

and

$$c^T_{mn} = \left(\frac{\partial \sigma_m}{\partial \epsilon_n}\right)_T, \tag{5.10}$$

where the superscripts S and T indicate the adiabatic and isothermal cases, respectively. In the usual situation, theoretical calculations are most conveniently carried out for the isothermal elastic constants, while ultrasonic measurements occur for the adiabatic regime, as will be shown in what follows.

An important concept from statistical mechanics [56, 57] is that of the partition function, Z, defined as

$$Z = \sum_s \exp\left(-\frac{E_s}{k_B T}\right). \tag{5.11}$$

The sum is over all the microstates, s, of the system with energy E_s, T is the temperature, and k_B is Boltzmann's constant. Generally these states are just the quantum mechanical states. The partition function may also be used classically in which case the sum is replaced by an integral over classical phase space. Many thermodynamic quantities are derivable from Z. For example, the Helmholtz free energy is obtained from the partition function by

$$F = -k_B T \ln Z. \tag{5.12}$$

A simple, but very useful, example is that of a harmonic oscillator for which the energies are, Equation 3.110,

$$E_s = (s + 1/2)\hbar\omega, \tag{5.13}$$

with $s = 0, 1, 2 \ldots$ The partition function is

$$Z = \sum_{s=0}^{\infty} \exp\left(-\frac{(s + 1/2)\hbar\omega}{k_B T}\right) = \exp\left(-\frac{\hbar\omega}{2k_B T}\right) \sum_{s=0}^{\infty} \left(\exp - \left(\frac{\hbar\omega}{k_B T}\right)\right)^s. \tag{5.14}$$

The power series can be summed, giving

$$Z = \exp\left(-\frac{\hbar\omega}{2k_B T}\right)\left(\frac{1}{1 - \exp - \left(\frac{\hbar\omega}{k_B T}\right)}\right) \tag{5.15}$$

Using Equation 5.12, F for the harmonic oscillator is

$$F_{H.osc} = \frac{\hbar\omega}{2} + k_B T \ln\left(1 - \exp\left(-\frac{\hbar\omega}{k_B T}\right)\right). \tag{5.16}$$

The first term in Equation 5.16 is due to the zero point motion. Equation 5.16 will be useful in the discussion of lattice vibrations.

5.3 Relation between Adiabatic and Isothermal Elastic Constants

The relation between adiabatic and isothermal elastic constants will now be developed. Equations 5.5 and 5.10 imply that [126]

$$d\sigma_m = \sum_{n=1}^{6} c^T_{mn} d\epsilon_n - f_m dT. \tag{5.17}$$

(Reference to Equation 5.3 shows that Equations 5.5 and 5.10 immediately follow from Equation 5.17.) Similarly [126],

$$dS = \sum_{n=1}^{6} f_n d\epsilon_n + (C_\epsilon/T)\, dT. \tag{5.18}$$

Equation 5.5 immediately follows from Equation 5.18; and

$$C_\epsilon = T\left(\frac{\partial S}{\partial T}\right)_\epsilon \tag{5.19}$$

where C_ϵ is the specific heat per unit volume at constant strain.

Setting $dS = 0$ and using dT from Equation 5.17 leads to

$$d\sigma_m = \sum_n c_{mn}^T d\epsilon_n + (T/C_\epsilon) f_m \sum_n f_n d\epsilon_n \tag{5.20}$$

for a constant entropy process. But, for a constant entropy process

$$d\sigma_m = \sum_n c_{mn}^S d\epsilon_n, \tag{5.21}$$

from Equation 5.9.

Comparing Equations 5.20 and 5.21 leads to

$$c_{mn}^S - c_{mn}^T = (T/C_\epsilon) f_m f_n. \tag{5.22}$$

It remains to clarify the meaning of the f_m. Setting $d\sigma_m = 0$ in Equation 5.17 (constant σ) and dividing by dT gives

$$f_m = \sum_n c_{mn}^T \alpha_n, \tag{5.23}$$

where

$$\alpha_n = \left(\frac{\partial \epsilon_n}{\partial T}\right)_\sigma . \tag{5.24}$$

Equations 5.22, 5.23, and 5.24 are general for the difference between adiabatic and isothermal elastic constants. Unless some of the components of the thermal expansion tensor [4], α_n, are negative (an unusual situation), the adiabatic elastic constants are larger than the isothermal ones. This effect may be understood qualitatively as follows. For the adiabatic case, a compressive stress produces a rise in temperature, which in turn produces a thermal expansion. Thus, the compressive strain is less than would be the case without the temperature rise. (The strains produced by thermal expansion and the compressive stress oppose each other.) By definition, there is no temperature rise in the isothermal case (any heat is conducted away from the volume under consideration), thus there is no thermal expansion and the compressive strain is more than in the adiabatic case. The result is that the material is stiffer in the adiabatic case.

For a more detailed discussion, the special case of cubic symmetry is now assumed. The thermal expansion tensor for cubic symmetry takes the form [4], [127] $\alpha_1 = \alpha_2 = \alpha_3 = \alpha; \alpha_4 = \alpha_5 = \alpha_6 = 0$. It follows that

$$c_{44}^S - c_{44}^T = 0 \tag{5.25}$$

and

$$c_{11}^S - c_{11}^T = c_{12}^S - c_{12}^T = \left(\beta B^T\right)^2 T/C_V \tag{5.26}$$

where $\beta = 3\alpha$ is the volume expansion coefficient, $B^T = \left(c_{11}^T + 2c_{12}^T\right)/3$ is the isothermal bulk modulus, and where, for cubic symmetry, $C_V = C_\epsilon$ [126]. The difference between that adiabatic and isothermal moduli is small in most cases, usually a few percent or so at room temperature.

Whether adiabatic or isothermal results are obtained depends on the experimental situation. From the study of the propagation of a temperature disturbance [56], during a time t an initial delta function disturbance will spread a distance l, where $l^2 \approx 4Dt$. D is the thermal diffusivity given by $D = \kappa/C_V$ and κ is the thermal conductivity. For propagation of ultrasonic waves, the relevant distance is $l = \lambda$ the wavelength. Comparing t to the period of the ultrasonic wave, then $t/\text{period} << 1$ is the isothermal case, while $t/\text{period} >> 1$ is the adiabatic case. This leads to

$$\frac{v^2}{4Df} << 1 \quad \text{isothermal;} \qquad \frac{v^2}{4Df} >> 1 \quad \text{adiabatic,} \tag{5.27}$$

where v is the ultrasonic velocity and f is the frequency of the wave. Substituting typical values into Equation 5.27 shows that ultrasonic measurements are in the adiabatic regime.

5.4 Elastic Constants and the Helmholtz Free Energy

In the discussion leading to Equations 2.62 and 2.63 the volume, V, was suppressed so that U and F in those equations, as also for other extensive thermodynamic properties, were per unit volume. In the discussion to follow, it is convenient to express the volume explicitly,

$$c_{mn}^S = \frac{1}{V}\left(\frac{\partial^2 U}{\partial \epsilon_m \partial \epsilon_n}\right)_S, \tag{5.28}$$

and

$$c_{mn}^T = \frac{1}{V}\left(\frac{\partial^2 F}{\partial \epsilon_m \partial \epsilon_n}\right)_T. \tag{5.29}$$

The typical approach is to calculate the isothermal elastic constants and obtain the adiabatic elastic constants from Equation 5.22; it is usually simpler to hold the temperature constant in taking derivatives, than it is to hold the entropy constant as in Equation 5.28.

In principle, a straightforward method to calculate the elastic constants of a material is now clear. Using Equation 5.11, calculate the partition function for the system of interest. (In some cases a classical approach involving an integral over classical phase space may be the appropriate way to calculate Z.) Next, calculate the Helmholtz free energy using the relation given by Equation 5.12. Assuming the strain-dependence of F is known, calculate the isothermal elastic constants with the use of Equation 5.29. Finally, use Equation 5.22 to obtain the adiabatic elastic constants. For relatively simple systems, *e.g.*, an isolated point defect, it is possible to carry out this prescription exactly as just described. In many cases, however, approximations are necessary.

5.5 *Ab Initio* Computations

Advances in computational solid-state physics, such as VASP [128] and Quantum Espresso [129], have made possible the *ab initio* calculation of many thermodynamic properties of crystals. These approaches use as input parameters the atomic number and the positions (lattice vectors) of the atoms in the unit cell [130].

It will prove useful to divide F into two parts: a part due to the static lattice; [43, 131] and, a part due to excitations

$$F = \phi_o + F_{exc}. \tag{5.30}$$

The static lattice energy, ϕ_o, means the potential energy in the absence of lattice vibrations, even zero-point motions. *At any temperature* the static lattice contribution to the Helmholtz free energy is the potential energy calculated for the atoms at rest at their mean equilibrium positions. The static lattice contribution depends implicitly on temperature, because ϕ_o depends on volume, which depends on temperature. The second term, F_{exc}, will typically be due to phonons, perhaps electrons, and other effects, depending on the material of interest.

The *ab initio* method is used with various levels of approximations. A brief description of some of these approaches will now be given.

5.5.1 Static Lattice Contribution at T = 0

At $T = 0$, $F = U$ and there is no difference between adiabatic and isothermal quantities. The *ab initio* methods enable the computation of ϕ_o at $T = 0$ for various atomic positions. These atomic positions can be the equilibrium positions, or the positions resulting from imposed strains. One approach is to compute ϕ_o for various computationally convenient strains, and determine the elastic constants by comparing the *ab initio* results with Equation 2.69, which gives the elastic energy associated with strains. (At $T = 0, U \neq \phi_o$, because of zero-point motions.)

The situation will be illustrated for the case of cubic symmetry.

1. First, let the strains be $\epsilon_1 = \epsilon_2 = \epsilon_3 = \delta$, $\epsilon_4 = \epsilon_5 = \epsilon_6 = 0$. For these particular strains, the *ab initio* approach is used to calculate ϕ_o/V as a function of δ. Next, using Equation 2.69 results in

$$\frac{\phi_o}{V} = \frac{3}{2}(c_{11} + 2c_{12})\,\delta^2 = \frac{9}{2}B\delta^2. \tag{5.31}$$

where ϕ_o has been taken as zero in the absence of strains. By fitting the *ab initio* results to a parabola, the numerical coefficient of δ^2 is found. Comparison with Equation 5.31 gives the numerical value of the elastic constant combination $c_{11} + 2c_{12}$, or equivalently, the bulk modulus B.

2. In a similar manner, choosing the strains as $\epsilon_1 = \epsilon_2 = \delta/2, \epsilon_3 = -\delta$, $\epsilon_4 = \epsilon_5 = \epsilon_6 = 0$, results in

$$\frac{\phi_o}{V} = \frac{3}{4}(c_{11} - c_{12})\,\delta^2 \tag{5.32}$$

Comparing with the *ab initio* calculation for this set of strains yields the elastic constant combination $c_{11} - c_{12}$.

3. Finally, choosing the strains as $\epsilon_1 = \epsilon_2 = \epsilon_3 = 0$, $\epsilon_4 = \epsilon_5 = \epsilon_6 = \delta$, gives,

$$\frac{\phi_o}{V} = \frac{3}{2}c_{44}\delta^2 \tag{5.33}$$

Comparison with the *ab initio* results determines c_{44}. Thus, the computation of ϕ_o/V for the three sets of strains determines the three independent elastic constants for cubic symmetry c_{11}, c_{12}, and c_{44}. The strains chosen are not the only ones possible; many other possibilities exist.

Ab initio calculations of the static lattice energy have produced good accuracy for the 0 K elastic constants in several cases [132, 133]. However, by definition, zero-point phonon vibrations are not included in the static lattice energy calculations. Zero-point effects are typically of the order of a few percent [134], or larger for light elements [135].

5.5.2 Temperature Dependence of Elastic Constants

In principle, the way to calculate the temperature dependence of the elastic constants is as outlined in Section 5.4. At a given temperature, find the Helmholtz free energy and compute the second derivatives of F with respect to the strains. Repeat the procedure at various temperatures. This procedure can be difficult to carry out in practice and various approximations are often used.

Effect of Lattice Vibrations

A treatment of lattice vibrations usually starts with the *harmonic approximation*. The harmonic approximation results from the the expansion of the potential energy in terms of displacements of atoms from their equilibrium positions and only keeping terms through second order in these displacements [136]. The vibrational effects can be described in terms of a set of independent oscillators of frequency ω_n. The quantized energy levels of these oscillators – phonons – are just those of harmonic oscillators. Thus, by Equations 5.16 and 5.30

$$F = \phi_o + \sum_{n=1}^{3N} \left[\frac{\hbar\omega_n}{2} + k_B T \ln \left(1 - \exp \left(-\frac{\hbar\omega_n}{k_B T} \right) \right) \right] \tag{5.34}$$

where ω_n is the frequency of the *nth* phonon mode, and the sum is over the $3N$ normal vibrational modes of the N atoms. Equation 5.34 may also be written as

$$F = \phi_o + \int_0^\infty \left(\frac{\hbar\omega}{2} + k_B T \ln \left(1 - \exp \left(-\frac{\hbar\omega}{k_B T} \right) \right) \right) g(\omega) d\omega \tag{5.35}$$

where $g(\omega)$ is the phonon density of states. In the harmonic approximation there is no thermal expansion [39]. With increasing temperature, the atoms vibrate at higher amplitudes, but the *average* position is unchanged. In addition, the elastic constants are independent of temperature in this approximation [136]. Higher order terms are needed for a more accurate description. These higher order terms are referred to as anharmonic effects.

The *quasi-harmonic approximation* [136] is frequently used to improve upon the harmonic approximation. In this approximation ω_n, or equivalently $g(\omega)$, is assumed to be volume dependent. This assumption is implemented by calculating $g(\omega)$ at various volumes [130, 135]. The result is that F is obtained as a function of V and T. However, the interest is in the *equilibrium* volume, denoted here by V_{eq}. Noting that $(\partial F/\partial V)_T = 0$ at equilibrium, V_{eq} is found by minimizing F – Equation 5.34 or Equation 5.35 – with respect to V at each temperature [130, 137, 138]. The result is $V_{eq}(T)$, which is equivalent to the thermal expansion. The free energy can then be written $F(T) = F(V_{eq}(T), T)$. The isothermal elastic constants are then calculated as

$$c_{mn}^T = \frac{1}{V_{eq}} \left(\frac{\partial^2 \phi_o(V_{eq}(T))}{\partial\epsilon_m \partial\epsilon_n} + \frac{\partial^2 \left[\sum_{n=1}^{3N} \left[\frac{\hbar\omega_n(V_{eq}(T))}{2} + k_B T \ln \left(1 - \exp \left(-\frac{\hbar\omega_n(V_{eq}(T))}{k_B T} \right) \right) \right] \right]}{\partial\epsilon_m \partial\epsilon_n} \right). \tag{5.36}$$

Equation 5.36 illustrates the method to calculate the elastic constants, including the effects of lattice vibrations, in the quasi-harmonic approximation. To summarize, ϕ_o and the ω_n are calculated as a function of volume. Next, the equilibrium volume, V_{eq} is found for various temperatures by minimizing F – Equation 5.34 – with respect to volume. Having found V_{eq} as a function of T, the implicit dependence of the elastic constants on T through the volume dependence can be removed, and the temperature dependence of the c_{mn}^T expressed by Equation 5.36.

The second term on the right-hand side of Equation 5.36 represents a formidable calculation in that the strain dependence of the ω_n must be calculated at various volumes. In some cases, this term is neglected [138, 139], and yet reasonable agreement with experimental results is obtained. The neglect of this term is equivalent to assuming that the temperature dependence of the elastic constants is essentially due to the temperature-induced volume expansion, not the direct temperature dependence expressed in Equation 5.36. There is other evidence that the primary cause of the temperature dependence of the elastic constants for many materials occurs indirectly through thermal expansion of the lattice [134].

Electronic Effects

For metals, the contribution of conduction electrons to the Helmholtz free energy should be included in Equations 5.34 and 5.35 [130, 138, 140]. For the electronic contribution this is conveniently accomplished by using the basic definition of the Helmholtz free energy,

$$F_e = \Delta U_e - S_e T. \tag{5.37}$$

The electronic thermal energy (total energy minus the $T = 0$ energy) is given by [39, 40]

$$\Delta U_e = \int_0^\infty n(\varepsilon) f_{FD}(\varepsilon, T)\varepsilon d\varepsilon - \int_0^{\varepsilon_F} n(\varepsilon)\varepsilon d\varepsilon, \tag{5.38}$$

where $n(\varepsilon)$ is electronic density of states at electron energy ε, and $f_{FD}(\varepsilon, T)$ is the Fermi-Dirac distribution function

$$f_{FD}(\varepsilon, T) = \frac{1}{\exp\left(\frac{\varepsilon - \mu(T)}{k_B T}\right) + 1} \tag{5.39}$$

with μ being the chemical potential. The chemical potential is determined by requiring[1]

$$N = \int_0^{\varepsilon_F} n(\varepsilon) d\varepsilon = \int_0^\infty n(\varepsilon) f_{FD}(\varepsilon, T) d\varepsilon \tag{5.40}$$

at each temperature T, where N is the number of conduction electrons.

Once $n(\varepsilon)$ and $\mu(T)$ are known it is possible to calculate the entropy, which is given by [138] and [141] (p. 161),

[1] The Fermi energy, ε_F, the highest occupied level at $T = 0$ is related to the chemical potential by $\varepsilon_F = \mu(0)$.

$$S_e = -k_B \int_0^{\infty} n(\varepsilon) \left[f_{FD}(\varepsilon,T) \ln(f_{FD}(\varepsilon,T)) + (1 - f_{FD}(\varepsilon,T)) \ln(1 - f_{FD}(\varepsilon,T)) \right] d\varepsilon \tag{5.41}$$

Notice that the entire procedure depends on the calculation of $n(\varepsilon)$, which in principle is accomplished with one of the *ab initio* methods for electronic structure calculations [142].

Including F_e, Equation 5.30 becomes[2]

$$F = \phi + F_{vib} + F_e \tag{5.42}$$

where F_{vib} is due to lattice vibrations, the second part of Equation 5.34. As was the case for the development of Equation 5.36, expansion of the lattice must be addressed. In principle $g(\omega)$, the phonon density of states, and $n(\varepsilon)$ the electronic density of states, are calculated for various volumes. As a result F is found as a function of V and T. As before, the interest is in the *equilibrium* volume, denoted here by V_{eq}. Because $(\partial F/\partial V)_T = 0$ at equilibrium, V_{eq} is found by minimizing F with respect to V at each temperature [130, 137, 138, 143]. The result is $V_{eq}(T)$, the thermal expansion [143]. Then the Helmholtz free energy can be expressed as [130]

$$F(T) = \phi((V_{eq}(T)) + F_{vib}(V_{eq}(T),T) + F_e(V_{eq}(T),T). \tag{5.43}$$

If it is possible to calculate $g(\omega)$ and $n(\epsilon)$ for various strains, then the elastic constants follow from

$$c_{ijkl} = \frac{1}{V_{eq}(T)} \left(\frac{\partial^2 F(T)}{\partial \epsilon_{ij} \partial \epsilon_{kl}} \right)_T . \tag{5.44}$$

Because this calculation is quite difficult, often the quasistatic approach is used where it is assumed that the temperature dependence is due primarily to the indirect effect on ϕ, *i.e.*, thermal expansion due to conduction electrons and thermal vibrations changes V_{eq} and hence causes ϕ to change; the explicit temperature dependence shown in F_{vib} and F_e is neglected. Such an approach gives reasonably accurate results [143].

5.6 Analytical Methods

Although the *ab initio* methods are very powerful and are performed with relatively few approximations, they can seem lacking in insight. The inputs are the atomic number and the lattice structure. The outputs are numerical values of the elastic constants, sometimes as a function of temperature. It is useful to have simple,

[2] The last term in Equation 5.38 is the kinetic energy of the conduction electrons at 0 K. By subtracting this last term in Equation 5.38, it is assumed that this energy is *included* in ϕ of Equation 5.42. This ground-state electronic energy term makes an appreciable contribution to the bulk modulus of metals [40].

idealized, models for which analytical treatments are possible. Such models often give results dependent on experimentally-determined parameters, such as the zero temperature elastic constants or Grüneisen parameters. However, these models may give other important information, *e.g.*, explicit temperature dependences. For example, it is useful to know under what conditions one may expect the elastic constants to depend on T, T^2, or T^4 in the ideal case. The extent to which the experimental results deviate from the idealized case may provide useful insight.

5.6.1 Static Lattic Contribution

As shown in Section 5.5.2, the static lattice contributes to the temperature dependence of the elastic constants indirectly through thermal expansion, indeed it appears to be the major contribution to the temperature dependence. In addition to its contribution to the temperature dependence, the pressure dependence of the elastic constants may also be understood through the volume dependence of ϕ_o [144]. Also, the (usually) anisotropic strains in deposited thin films affect the elastic constants through the strain dependence of ϕ_o [145]. For all three problems – pressure dependence of the elastic constants, effects of strains in thin films on elastic constants, and the effect of the static lattice on the temperature dependence of the elastic constants – it is convenient to divide the strain into two parts. There is an initial strain, which may be *relatively* large; and there is an additional, superimposed strain, due, *e.g.*, to ultrasonic waves, which is relatively small. The response of the lattice to the small strains differ in the presence of the initial strain, as compared to the case with no initial strain. The objective in this discussion is to express the elastic constants measured *after* the initial strain in terms of parameters of the initial state. Thus, one application would be to express the elastic constants of a film subjected to relatively large strains produced by a substrate, in terms of second, and higher-order elastic constants of the *unstrained* film. Another application would be to express the elastic constants measured after a thermal expansion in terms of parameters of the lattice *before* the expansion. In this way the implicit temperature dependence due to the static lattice expansion is found.

Relevant Strains

As discussed in Section 2.1, strains describe the distortion of a material from an initial configuration to a final configuration. For the problem to be addressed in the following discussion there are at least two strains: an initial one to produce a strained state; and, an additional one to measure the elastic response relative to the strained state. It turns out to be convenient to introduce a *third* strain to facilitate the discussion. These three strains will be explored in what follows, although not in the order just mentioned.

x $\Rightarrow$ **x**′ It may be useful at this point to review briefly the *general discussion* of Section 2.1 and adapt it to the present situation. Following Equation 2.1 we denote the displacement produced by a strain as

$$u_i = x_i' - x_i, \tag{5.45}$$

where x_i represents the original position of a point before the deformation and x_i' the final position.

For the present case the strains will be small, $<< 1$, but not necessarily infinitesimal. Equation 2.8 for finite strains will be needed, which is repeated here

$$\epsilon_{ij} = \frac{1}{2}\left(\frac{\partial u_i}{\partial x_j} + \frac{\partial u_j}{\partial x_i} + \sum_l \frac{\partial u_l}{\partial x_j}\frac{\partial u_l}{\partial x_i}\right). \tag{5.46}$$

In the treatment following Equation 2.8, the last term in Equation 5.46 was neglected, but that term is important for the present discussion.

An alternate expression for ϵ_{ij} will be helpful. Defining

$$\alpha_{ij} = \frac{\partial x_i'}{\partial x_j} \tag{5.47}$$

and

$$u_{ij} = \frac{\partial u_i}{\partial x_j}, \tag{5.48}$$

it follows from Equation 5.45 that

$$u_{ij} = \alpha_{ij} - \delta_{ij}. \tag{5.49}$$

Substituting Equation 5.49 into Equation 5.46 gives [144]

$$\epsilon_{ij} = \frac{1}{2}\left(\sum_{l=1}^{3}\left(\alpha_{li}\alpha_{lj}\right) - \delta_{ij}\right). \tag{5.50}$$

This alternate form for the strain will prove useful shortly.

$\bar{\mathbf{x}} \Rightarrow \mathbf{x}'$ It is now convenient to introduce a *special* initial reference set of coordinates denoted by $\bar{x}_i$. This initial configuration might correspond to zero strain or some other special condition such as zero temperature. The objective will be to use a Maclaurin expansion of thermodynamic potentials about the initial configurations $\bar{x}_i$, to express the elastic response relative to the coordinates x_i of Equation 5.45. The coefficients in the expansion will involve higher-order elastic constants relative to the initial configuration $\bar{x}_i$ [144]. Examples have already been mentioned: the elastic constants under pressure in terms of the zero pressure higher-order constants; the elastic constants in a film strained by the substrate in terms of the unstrained film; and the elastic constants at higher temperatures in terms of the zero-temperature higher-order elastic constants. In the last case, the strain producing a configuration change from $\bar{x}_I$ to x_i is produced by thermal expansion.

In keeping with the previous notation the strains relative to the initial configuration $\bar{x}_i$ are defined as

$$\bar{\epsilon}_{ij} = \frac{1}{2}\left(\sum_l \left(\bar{\alpha}_{li}\bar{\alpha}_{lj}\right) - \delta_{ij}\right). \tag{5.51}$$

with

$$\bar{\alpha}_{ij} = \frac{\partial x'_i}{\partial \bar{x}_j} \tag{5.52}$$

As Equation 5.52 implies, $\bar{\epsilon}_{ij}$ represent the strains from $\bar{x}$ to x' ($\bar{x}$ the original, special, configuration, x' the final configuration).

It will prove useful to have a relation between ϵ_{ij} and $\bar{\epsilon}_{ij}$, in particular, $\partial\bar{\epsilon}_{ij}/\partial\epsilon_{ij}$ is needed. It is possible to transform between the two strains using the chain rule for differentiation

$$\bar{\alpha}_{ij} = \sum_{k=1}^{3} \alpha_{ik} a_{kj} \tag{5.53}$$

where

$$a_{kj} = \frac{\partial x_k}{\partial \bar{x}_j}. \tag{5.54}$$

using Equations 5.50 to 5.54 results in

$$\left(\bar{\epsilon}_{ij} + \frac{1}{2}\delta_{ij}\right) = \sum_{rs}\left(\epsilon_{rs} + \frac{1}{2}\delta_{rs}\right) a_{ri}a_{sj}. \tag{5.55}$$

The desired connection between the two strains follows [144],

$$\frac{\partial\bar{\epsilon}_{ij}}{\partial\epsilon_{rs}} = a_{ri}a_{sj}. \tag{5.56}$$

$\bar{\mathbf{x}} \Rightarrow \mathbf{x}$ Finally, the arguments in the following Section are a bit subtle, and it will be useful to define a third set of strains at this point, e_{ij} (not ϵ_{ij}), involving the transformation from $\bar{x}$ to x [7]. Folllowing the above definitions and development, and using Equation 5.54

$$e_{ij} = \frac{1}{2}\left(\sum_l \left(a_{li}a_{ij}\right) - \delta_{ij}\right). \tag{5.57}$$

Expansion of F in Terms of the $\bar{\epsilon}_{ij}$

The Helmholtz free energy will be expanded about the initial configuration, $\bar{x}_i$, in terms of the strains $\bar{\epsilon}_{ij}$ [144]. As already mentioned, these strains could be due to the application of external forces such as hydrostatic pressure, or perhaps due to forces exerted by a substrate on a thin film. The present expansion will be applied to the case in which these strains are due to thermal expansion of the material. In this manner, the static lattice contribution to the temperature dependence of the elastic constants is obtained.

The general expression for a Taylor series for multiple variables is [9]

$$f(x_1 \ldots x_d) = \left[\sum_{n=0}^{\infty} \frac{1}{n!} \left[\sum_{i=1}^{d} (x_i - x_{io}) \frac{\partial}{\partial x_i} \right]^n f(x_i \ldots x_d)_{x_i = x_{io}} \right] \tag{5.58}$$

where there are d variables, and the function $f(x_i)$ is to be expanded about the positions $x_i = x_{io}$. For the present case the variables are the nine $\bar{\epsilon}_{ij}$, and the function is $F(\bar{\epsilon}_{ij})$, which is to be expanded about $\bar{\epsilon}_{ij} = 0$. The result is

$$\begin{aligned} F(\bar{\epsilon}_{ij}) = F(0) &+ \frac{1}{2} \sum_{ijkl} \left(\frac{\partial^2 F}{\partial \bar{\epsilon}_{ij} \partial \bar{\epsilon}_{kl}} \right)_{\bar{\epsilon}=0} \bar{\epsilon}_{ij} \bar{\epsilon}_{kl} \\ &+ \frac{1}{3!} \sum_{ijklmn} \left(\frac{\partial^3 F}{\partial \bar{\epsilon}_{ij} \partial \bar{\epsilon}_{kl} \partial \bar{\epsilon}_{mn}} \right)_{\bar{\epsilon}=0} \bar{\epsilon}_{ij} \bar{\epsilon}_{kl} \bar{\epsilon}_{mn} \\ &+ \frac{1}{4!} \sum_{ijklmnpq} \left(\frac{\partial^4 F}{\partial \bar{\epsilon}_{ij} \partial \bar{\epsilon}_{kl} \partial \bar{\epsilon}_{mn} \partial \bar{\epsilon}_{pq}} \right)_{\bar{\epsilon}=0} \bar{\epsilon}_{ij} \bar{\epsilon}_{kl} \bar{\epsilon}_{mn} \bar{\epsilon}_{pq} + \cdots \end{aligned} \tag{5.59}$$

where, as usual, each index $i, j, k, l \ldots$ ranges from 1 to 3. First derivative terms would correspond to initial stresses, Equation 2.61, which are zero in the present case.

Equation 5.59 can be rewritten in terms of, the second, third, and fourth order isothermal elastic constants [7],

$$\begin{aligned} \bar{c}_{ijkl} &= \frac{1}{V_0} \left(\frac{\partial^2 F}{\partial \bar{\epsilon}_{ij} \partial \bar{\epsilon}_{kl}} \right)_{T,\bar{\epsilon}=0} \\ \bar{c}_{ijklmn} &= \frac{1}{V_0} \left(\frac{\partial^3 F}{\partial \bar{\epsilon}_{ij} \partial \bar{\epsilon}_{kl} \partial \bar{\epsilon}_{mn}} \right)_{T,\bar{\epsilon}=0} \\ \bar{c}_{ijklmnpq} &= \frac{1}{V_0} \left(\frac{\partial^4 F}{\partial \bar{\epsilon}_{ij} \partial \bar{\epsilon}_{kl} \partial \bar{\epsilon}_{mn} \partial \bar{\epsilon}_{pq}} \right)_{T,\bar{\epsilon}=0}, \end{aligned} \tag{5.60}$$

giving

$$\begin{aligned} \frac{F(\bar{\epsilon}_{ij})}{V_0} = \frac{F(0)}{V_0} &+ \frac{1}{2} \sum_{ijkl} \bar{c}_{ijkl} \bar{\epsilon}_{ij} \bar{\epsilon}_{kl} \\ &+ \frac{1}{3!} \sum_{ijklmn} \bar{c}_{ijklmn} \bar{\epsilon}_{ij} \bar{\epsilon}_{kl} \bar{\epsilon}_{mn} \\ &+ \frac{1}{4!} \sum_{ijklmnpq} \bar{c}_{ijklmnpq} \bar{\epsilon}_{ij} \bar{\epsilon}_{kl} \bar{\epsilon}_{mn} \bar{\epsilon}_{pq} + \cdots \end{aligned} \tag{5.61}$$

Note that the condensed (Voigt) notation is *not* used in these expressions. The derivatives are to be evaluated at $\bar{\epsilon}_{ij} = 0$. The bar over the expression for the elastic constants, *e.g.*, $\bar{c}_{ijkl}$, indicates that these equations represent the second, third, and fourth-order elastic constants at the *initial configuration*, $\bar{x}_i$. The initial configuration should be defined for each application. Being the derivatives of F, these are the isothermal constants. V_0 is the volume of a given mass of the material at the original configuration. Below, V_1 will represent the volume of this same mass of material at the strained configuration.

The indices in Equation 5.61 each run from 1 to 3. Equation 5.61 gives the Helmholtz free energy *for the strained lattice, i.e.* $\bar{\epsilon}_{ij}$, in terms of higher order elastic constants *at the initial configuration.*

Elastic Constants of the Strained Lattice

It is desired to know the second-order elastic constants for the *strained lattice*. Thus, it is just necessary to take second derivatives of $F(\bar{\epsilon}_{ij})$ with respect to ϵ_{ij}, (not $\bar{\epsilon}_{ij}$).

Using Equations 5.29, 5.56, and the chain rule for differentiation, the isothermal second-order elastic constants of the strained lattice are

$$c_{ijkl} = \frac{1}{V_1}\left(\frac{\partial F(\bar{\epsilon})}{\partial \epsilon_{ij}\partial \epsilon_{kl}}\right)_{\epsilon=0} = \frac{1}{V_1}\sum_{mnpq} a_{im}a_{jn}a_{kp}a_{lq}\left(\frac{\partial^2 F(\bar{\epsilon})}{\partial \bar{\epsilon}_{mn}\partial \bar{\epsilon}_{pq}}\right)_{\epsilon=0}, \tag{5.62}$$

where V_1 is now used rather than V_o. The derivatives in Equation 5.62 are to be evaluated at $\epsilon = 0$ (not $\bar{\epsilon} = 0$).

Using Equation 5.61 the final derivative on the right-hand side of Equation 5.62 may be computed. First, however, Equation 5.61 will be rewritten to correspond closely to the notation in Equation 5.62.

$$\begin{aligned} F(\bar{\epsilon}) = F(0) + V_0 \Bigg[& \frac{1}{2}\sum_{mnpq} \bar{c}_{mnpq}\bar{\epsilon}_{mn}\bar{\epsilon}_{pq} \\ & + \frac{1}{3!}\sum_{mnpqrs} \bar{c}_{mnpqrs}\bar{\epsilon}_{mn}\bar{\epsilon}_{pq}\bar{\epsilon}_{rs} \\ & + \frac{1}{4!}\sum_{mnpqrstu} \bar{c}_{mnpqrstu}\bar{\epsilon}_{mn}\bar{\epsilon}_{pq}\bar{\epsilon}_{rs}\bar{\epsilon}_{tu} + \cdots \Bigg]. \end{aligned} \tag{5.63}$$

Finally,[3]

$$c_{ijkl} = \frac{V_0}{V_1}\left[\sum_{mnpq} a_{im}a_{jn}a_{kp}a_{lq}\left[\bar{c}_{mnpq} + \sum_{rs}\bar{c}_{mnpqrs}e_{rs} + \frac{1}{2}\sum_{rstu}\bar{c}_{mnpqrstu}e_{rs}e_{tu}\right] + \cdots\right]. \tag{5.64}$$

[3] Equation 5.64 is the same as equation 3.10 of Ref. [144]. An abbreviated version is given in Ref. [145].

The derivatives in Equation 5.62 are evaluated at $\epsilon_{ij} = 0$ in which case the $\bar{\epsilon}_{ij} \rightarrow e_{ij}$, so it is the latter strains that appear in Equation 5.64. (In practice the difference between $\bar{\epsilon}_{ij}$ and e_{ij} is quite small, $<< \bar{\epsilon}_{ij}$.)

It is important to keep in mind a few key points. Equation 5.64 gives the elastic constants of the strained lattice, c_{ijkl}, in terms of the second and higher-order elastic constants, $\bar{c}_{mnpqrs}$, etc., *evaluated for the initial configurations, usually the unstrained lattice.* It seems worth repeating that Equation 5.64 gives the static lattice contribution to the temperature dependence of the elastic constants, or to the pressure dependence of elastic constants, or to the elastic constant of strained films. It is only necessary to use the appropriate e_{ij} for each case. This equation looks exceptionally complicated, but great simplifications occur for high symmetries.

Equation 5.64 has been derived using derivatives of F, so all elastic constants appearing in the equation are for the *isothermal* case. Exactly the same type of equation is found using derivatives of U instead, in which case all the elastic constants in the equation are for the *adiabatic case* [144].

Applicaton to Cubic Symmetry

Thermal expansion is now considered. The coordinates of a particular material point at the initial temperature are $(\bar{x}_1, \bar{x}_2, \bar{x}_3)$. The coordinates of the same material point at the new temperature are (x_1, x_2, x_3). If the material is cubic throughout the process, then

$$x_i = (1+\xi)\bar{x}_i \tag{5.65}$$

giving

$$a_{ij} = (1+\xi)\delta_{ij} \tag{5.66}$$

and

$$V_1 \approx (1+\xi)^3 V_0 \tag{5.67}$$

where ξ is the fractional change in length, along a particular direction. Using Equations 5.66 and 5.67, Equation 5.64 becomes

$$c_{ijkl} = (1+\xi)\left[\bar{c}_{ijkl} + \sum_{rs} \bar{c}_{ijklrs} e_{rs} + \frac{1}{2}\sum_{rstu} \bar{c}_{ijklrstu} e_{rs} e_{tu}\right] + .. \tag{5.68}$$

Next, Equations 5.51, 5.52, and 5.65 give

$$e_{rs} = \left[\xi + \frac{1}{2}\xi^2\right]\delta_{rs} \tag{5.69}$$

for the thermal strain. Combining Equations 5.68 and 5.69 yields,

$$c_{ijkl} = (1+\xi)\left[\bar{c}_{ijkl} + \left[\xi + \frac{1}{2}\xi^2\right]\sum_{r} \bar{c}_{ijklrr} + \frac{1}{2}\left[\xi + \frac{1}{2}\xi^2\right]^2 \sum_{rt} \bar{c}_{ijklrrtt}\right]. \tag{5.70}$$

Recall that ξ is defined by Equation 5.65.

Static Lattice Contribution to the Temperature Dependence, the Quasistatic Approximation

Equation 5.70 expresses the **static lattice** *contribution to the second-order elastic constant, after expansion, in terms of the second-, third-, and fourth-order elastic constants before the expansion,* $\bar{c}_{ijkl}$, $\bar{c}_{ijklrr}$, *and* $\bar{c}_{ijklrrtt}$. Equation 5.65 is for cubic symmetry, so Equation 5.70 is also so restricted, although the results are easily generalized to other symmetries. This result may be applied to several different situations, as discussed in Section 5.6.1. For the present case, the expansion is taken to be due to a temperature increase, in which case Equation 5.70 gives the temperature dependence of the elastic constants in the static lattice approximation. The thermal expansion is normally small, $\xi << 1$, so it is sufficient to keep only first-order terms in ξ,

$$c_{ijkl} \simeq \bar{c}_{ijkl} + \xi \left[\bar{c}_{ijkl} + \sum_r \bar{c}_{ijklrr} \right]. \tag{5.71}$$

Equation 5.71 has been given previously by Garber and Granato [131].

There are six independent third-order elastic constants for cubic symmetry [146]. Using this simplification plus the symmetries given by Birch [146], Equation 5.71 gives for cubic symmetry [144]

$$\begin{aligned} c_{11} &\simeq \bar{c}_{11} + \xi\,(\bar{c}_{11} + \bar{c}_{111} + 2\bar{c}_{112}) \\ c_{12} &\simeq \bar{c}_{12} + \xi\,(\bar{c}_{12} + \bar{c}_{123} + 2\bar{c}_{112}) \\ c_{44} &\simeq \bar{c}_{44} + \xi\,(\bar{c}_{44} + \bar{c}_{144} + 2\bar{c}_{166}). \end{aligned} \tag{5.72}$$

Equatons 5.72 give the *static lattice* contribution to the elastic constants of cubic crystals, which have undergone a strain ξ, in terms of second- and third-order elastic constants of the lattice in the initial configuration (unstrained). The following application is to the case where the strains are due to thermal expansion. It has been suggested [138, 139, 143] that this contribution, *quasistatic*, is the dominant contribution to the temperature dependence of the c_{ij}. This suggestion is tested in Appendix E by applying Equation 5.72 to specific cases.

5.6.2 *Quasiharmonic Treatment of Thermal Vibrations*

In the quasistatic model, the temperature dependence of the elastic constants arises because thermal expansion changes the static lattice contribution to the elastic constants. Appendix E compares the quasistatic model to experimental results by using experimental values for the thermal expansion. In the present section the *full* contribution of the vibrations to the temperature dependence is explored. In general the equations derived are rather complex, and simple results are obtained only in limiting cases. The present discussion is closely related to Ref. [131].

Using Equations 5.29 and 5.34, the isothermal, second order elastic constants may be written

$$c_{ijkl}^{T} = \tilde{c}_{ijkl}^{T} + \frac{1}{V}\sum_{n=1}^{3N}\left[\left(\gamma_n^{ij}\gamma_n^{kl} - \frac{\partial \gamma_n^{ij}}{\partial \epsilon_{kl}}\right)u_n - \gamma_n^{ij}\gamma_n^{kl}c_n T\right]. \tag{5.73}$$

Equation 5.73 introduces several new parameters which will now be defined.

$$\tilde{c}_{ijkl}^{T} = \frac{1}{V}\left(\frac{\partial^2 \phi_o}{\partial \epsilon_{ij}\partial \epsilon_{kl}}\right). \tag{5.74}$$

The notation $\tilde{c}_{ijkl}$ refers to the part of the elastic constant due to ϕ_o, the *static* lattice contribution. This contribution is static in the sense that the explicit effects of *both* zero-point motion *and* thermal vibrations appearing as the second term in Equation 5.73 are *not* included in $\tilde{c}_{ijkl}$; but, these effects do influence $\tilde{c}_{ijkl}$ indirectly through the vibration-induced lattice expansion, which will normally affect ϕ_o. These effects will be addressed below.

The vibrational energy for mode n, Equation 3.111, is repeated here[4]

$$u_n = \left(\langle s_n \rangle + \frac{1}{2}\right)(\hbar\omega_n), \tag{5.75}$$

where $\langle s_n \rangle$, the thermal average number of quantum excitations (phonons) in mode n, is given by the Planck distribution function, Equation 3.112,

$$\langle s_n \rangle = \frac{1}{\exp(\hbar\omega_n/k_B T) - 1}. \tag{5.76}$$

c_n is the heat capacity for mode n,

$$c_n = \frac{\partial u_n}{\partial T} = \frac{(\hbar\omega_n)^2}{k_B T^2}\frac{\exp\frac{\hbar\omega_n}{k_B T}}{\left(\exp\frac{\hbar\omega_n}{k_B T} - 1\right)^2}. \tag{5.77}$$

Finally, the dependence of the phonon frequencies on the strains is approximated by the isothermal, generalized-mode Grüneisen parameter,

$$\gamma_n^{ij} = -\frac{1}{\omega_n}\left(\frac{\partial \omega_n}{\partial \epsilon_{ij}}\right)_T. \tag{5.78}$$

Equation 5.71 is readily adapted to the present situation. That equation, for a cubic material, gives the effects of a lattice expansion on the static lattice contribution to the elastic constants. The results are expressed in terms of the expansion

[4] It is important to remember that each mode n is specified by both the K value and the polarization. There are three modes with (generally) different frequencies for each K value.

relative to an initial reference state. For the present case the reference state is taken as 0 K *in the absence of zero-point motion.*

$$\tilde{c}^T_{ijkl} \simeq \tilde{c}^o_{ijkl} + \tilde{\xi}^o \left[\tilde{c}^o_{ijkl} + \sum_r \tilde{c}^o_{ijklrr} \right]. \tag{5.79}$$

Here, $\tilde{c}^o_{ijkl}$ is the static lattice value at 0 K and does not include zero-point motion effects, while $\tilde{c}^T_{ijkl}$ is the static lattice contribution to the elastic constants at strain $\tilde{\xi}^o$, produced by thermal expansion and zero-point motion. Thus, $\tilde{c}^T_{ijkl}$ is temperature dependent.

It remains to determine ξ for the present situation, *including expansion due to zero-point motion.* (In Appendix E, experimental values were used for ξ. In addition, only thermal expansion was involved, not zero-point motion induced expansion.)

Using the fact that $(\partial F/\partial \epsilon_{ij})_T = 0$ at equilibrium for no external stresses, Equations 5.34, 3.111, 3.112, and 5.78 yield,

$$\left(\frac{\partial \phi_o}{\partial \epsilon_{ij}}\right)_T = \sum_n \gamma_n^{ij} u_n. \tag{5.80}$$

The next step is to expand the left-hand side of Equation 5.80 in term of the strains relative to the the state at 0 K *without zero-point motion,*

$$\left(\frac{\partial \phi_o}{\partial \epsilon_{ij}}\right)_T \simeq \left(\frac{\partial \phi_o}{\partial \epsilon_{ij}}\right)^o + \sum_{kl} \left(\frac{\partial^2 \phi_o}{\partial \epsilon_{kl} \partial \epsilon_{ij}}\right)^o \tilde{\epsilon}^o_{kl}. \tag{5.81}$$

The first term on the right-hand side of Equation 5.81 is zero at equilibrium. Combining Equations 5.80 and 5.81 gives

$$\sum_{kl} \left(\frac{\partial^2 \phi_o}{\partial \epsilon_{kl} \partial \epsilon_{ij}}\right)^o \tilde{\epsilon}^o_{kl} = \sum_n \gamma_n^{ij} u_n. \tag{5.82}$$

Simplifications are possible for cubic symmetry. By Equation 5.69, $\tilde{\epsilon}^o_{kl} = \tilde{\xi}^o \delta_{kl}$ to first order in $\tilde{\xi}^o$. (The superscript o indicates the reference state is at 0 K; the tilde indicates the reference state is without zero-point motion.) From Refs. [147, 148, 149],

$$\gamma_n = -\frac{V}{\omega_n}\left(\frac{\partial \omega_n}{\partial V}\right)_T = \frac{1}{3}\left(\gamma_n^{11} + \gamma_n^{22} + \gamma_n^{33}\right). \tag{5.83}$$

Using

$$\frac{1}{\tilde{V}^o}\left(\frac{\partial^2 \phi_o}{\partial \epsilon_{kl} \partial \epsilon_{ij}}\right)^o = \tilde{c}^o_{ijkl}, \tag{5.84}$$

Equation 5.83, and taking into account that many elastic constants are zero for cubic symmetry, Equation 5.82 gives

$$\tilde{\xi}^o = \frac{1}{\tilde{V}^o} \frac{1}{\tilde{c}^o_{11} + 2\tilde{c}^o_{12}} \sum_n \gamma_n u_n, \tag{5.85}$$

where $\tilde{\xi}^o$ contains the expansion due to zero point motion. The energy u contains the explicit temperature dependence. Combining 5.79 and 5.85 gives the contribution of the static lattice, including the effect of lattice expansion,

$$\tilde{c}^T_{ijkl} = \tilde{c}^o_{ijkl} + \left[\frac{1}{(\tilde{c}^o_{11} + 2\tilde{c}^o_{12})\tilde{V}^o}\sum_n \gamma_n u_n\right]\left[\tilde{c}^o_{ijkl} + \sum_r \tilde{c}^o_{ijklrr}\right] \tag{5.86}$$

Finally, combining Equations 5.73 and 5.86 gives the full temperature dependence in the quasiharmonic approximation [131],

$$\begin{aligned} c^T_{ijkl}(T) = \tilde{c}^o_{ijkl} &+ \left[\frac{1}{(\tilde{c}^o_{11} + 2\tilde{c}^o_{12})\tilde{V}^o}\sum_n \gamma_n u_n\right]\left[\tilde{c}^o_{ijkl} + \sum_r \tilde{c}^o_{ijklrr}\right] \\ &+ \frac{1}{\tilde{V}^o}\sum_n\left[\left(\gamma^{ij}_n\gamma^{kl}_n - \frac{\partial \gamma^{ij}_n}{\partial \epsilon_{kl}}\right)u_n - \gamma^{kl}_n\gamma^{ij}_n c_n T\right]. \end{aligned} \tag{5.87}$$

In the spirit of the quasi-harmonic approximations, the frequencies ω_n are for oscillations about the static lattice and are temperature independent. This means that u_n and c_n are evaluated for such frequencies. The Grüneisen parameters and their strain derivatives are taken as temperature independent. Thus, the temperature dependence is given by the explicit T in Equation 5.87 (which originates in Equation 5.73) and the temperature dependences of u_n and c_n. Everything else in the equation is based on static lattice parameters. As the notation is already rather cumbersome, no special notation will be used for these conditions on u_n, c_n, and γ_n.

Equation 5.26 may be applied specifically to the present case to obtain the adiabatic elastic constants. The linear thermal expansion coefficient, α, is found from Equation 5.85

$$\alpha = \frac{d\tilde{\xi}^o}{dT} = \frac{1}{\tilde{V}^o(\tilde{c}^o_{11} + 2\tilde{c}^o_{12})}\sum_n \gamma_n c_n = \frac{1}{3\tilde{V}^o B^T}\sum_n \gamma_n c_n. \tag{5.88}$$

Combining Equations 5.26 and 5.88, and noting that $\beta = 3\alpha$, Equation 5.26, results in

$$c^S_{11} - c^T_{11} = c^S_{12} - c^T_{12} = \frac{T\left(\sum_n \gamma_n c_n\right)^2}{\tilde{V}^o \sum_n c_n} \tag{5.89}$$

with $C_V = \sum_n c_n/V^o$ as has been found previously [131].

Within the quasi-harmonic approximation, Equations 5.87 and 5.89 provide a complete description of the temperature dependence of the elastic constants due to thermal vibrations. A general solution is difficult because of the sums. To evaluate the sums it is necessary to know the normal mode frequencies ω_n, and their dependence on wave number $k_n = 2\pi/\lambda_n$ (λ_n is the wavelength associated with ω_n) for all directions in the crystal. Also, it is necessary to find the Grüneisen parameters, *i.e.*, the dependence of the ω_n on the strains. Considerable insight is gained by treating special cases, which is the approach used here.

High-Temperature Limit

The expression for u_n can be simplified in the high temperature limit, which is $k_B T >> \hbar\omega$ for all frequencies.

$$u_n = \hbar\omega_n \left(\frac{1}{2} + \frac{1}{\exp \frac{\hbar\omega_n}{k_B T} - 1} \right) \simeq \hbar\omega_n \left(\frac{1}{2} + \frac{1}{\frac{\hbar\omega_n}{k_B T} + \frac{1}{2} \left(\frac{\hbar\omega}{k_B T} \right)^2 \cdots} \right)$$

$$= \hbar\omega_n \left(\frac{1}{2} + \frac{1}{\left(\frac{\hbar\omega_n}{k_B T} \right) \left(1 + \frac{1}{2} \left(\frac{\hbar\omega_n}{k_B T} \right) \right)} \right) \simeq \hbar\omega_n \left(\frac{1}{2} + \frac{k_B T}{\hbar\omega_n} \left(1 - \frac{1}{2} \frac{\hbar\omega_n}{k_B T} \right) \right) = k_B T. \tag{5.90}$$

Notice how the factor $\left(\frac{1}{2}\right)\left(\frac{\hbar\omega_n}{k_B T}\right)^2$ in the expansion of the exponential exactly cancels the zero-point motion contribution [40]. The contribution of each mode n to the specific heat is just k_B in the high temperature limit, the classical result.

Applying Equation 5.90 to Equation 5.87 results in

$$c^T_{ijkl}(T) = \tilde{c}^o_{ijkl} - AT \tag{5.91}$$

with

$$A = -\frac{k_B}{\tilde{V}^o} \left(\frac{\left(\tilde{c}^o_{ijkl} + \sum_r \tilde{c}^o_{ijklrr} \right)}{\left(\tilde{c}^o_{11} + 2\tilde{c}^o_{12} \right)} \sum_n \gamma_n - \sum_n \frac{\partial \gamma^{ij}_n}{\partial \epsilon_{kl}} \right). \tag{5.92}$$

for the high-temperature limit. The parameter A is a constant in the quasiharmonic approximation, independent of T. As demonstrated in Appendix E the dominant contribution to the temperature dependence is the first term on the right-hand side of Equation 5.92, and that term gives a negative contribution in the usual case. Thus, A is normally positive.

Equation 5.91 shows two important results for the temperature dependence of the elastic constants produced by thermal lattice vibrations, in the high temperature limit. (1) The elastic constants decrease linearly with temperature in this range. This result is rather general [131]. (2) Further, the extrapolation of the high-temperature region to zero temperature gives $\tilde{c}^o_{ijkl}$, the completely static lattice result [131] (zero temperature and no zero-point motion effect).

Zero-Temperature Results

At $T = 0$ Equation 5.87 becomes

$$c^T_{ijkl}(0) = \tilde{c}^o_{ijkl} + \left[\frac{1}{(\tilde{c}^o_{11} + 2\tilde{c}^o_{12})\tilde{V}^o} \sum_n \gamma_n \hbar\omega_n/2 \right] \left[\tilde{c}^o_{ijkl} + \sum_r \tilde{c}^o_{ijklrr} \right]$$
$$+ \frac{1}{\tilde{V}^o} \sum_n \left[\left(\gamma^{ij}_n \gamma^{kl}_n - \frac{\partial \gamma^{ij}_n}{\partial \epsilon_{kl}} \right) \hbar\omega_n/2 \right]. \tag{5.93}$$

The terms on the right-hand side of Equation 5.93 are readily interpreted. The first term has been mentioned several times. It is the contribution from the potential energy the collection of atoms *would* have at their static $T = 0$, positions in the absence of zero-point motion. The second term includes the effect of the lattice expansion due to zero-point motion. The last term is a dynamic effect due to zero-point motion. *Thus, the difference between linear results extrapolated to* 0 *K, Equation 5.91, and the actual* $T = 0$ *values, Equation 5.93, is entirely due to zero-point motion effects.*

Low-Temperature Limit

Using Equation 5.93 it is now possible [150] to rewrite Equation 5.87 in a manner that is more convenient for the present purpose.

$$c^T_{ijkl}(T) = c^T_{ijkl}(0) + \left[\frac{1}{(\tilde{c}^o_{11} + 2\tilde{c}^o_{12})\tilde{V}^o}\sum_n \gamma_n\left(u_n - \frac{\hbar\omega_n}{2}\right)\right]\left[\tilde{c}^o_{ijkl} + \sum_r \tilde{c}^o_{ijklrr}\right] + \frac{1}{\tilde{V}^o}\sum_n\left[\left(\gamma^{ij}_n\gamma^{kl}_n - \frac{\partial\gamma^{ij}_n}{\partial\epsilon_{kl}}\right)(u_n - \hbar\omega_n/2) - \gamma^{kl}_n\gamma^{ij}_n c_n T\right]. \tag{5.94}$$

Equation 5.94 is not restricted in temperature range, within the quasiharmonic approach.

The next step is to consider approximations valid in the low-temperature limit. The sums in Equation 5.94 are difficult to evaluate in part because u_n and c_n depend on frequency, while γ_n, γ^{ij}_n, and $\partial\gamma^{ij}_n/\partial\epsilon_{kl}$ may depend on both the frequency and the direction of propagation of the thermally-excited lattice waves.

To proceed further, the Debye model will be used (Section 3.3), which is an accurate description for many condensed-matter properties in the low-temperature limit. In the low-temperature limit, the long-wavelength waves make the primary contribution to u_n and c_n. In this regime, the neglect of any wavelength (frequency) dependence of the Grüneisen parameters is expected to be valid [131, 148], and *this assumption is made in what follows. If* elastic isotropy is also assumed, the Grüneisen parameter terms are constant within the stated assumptions, and the sums are readily evaluated in the Debye approximation [40, 56]. The thermal energy term $\sum_n (u_n - \hbar\omega_n/2)$ varies as T^4, and specific heat term $\sum_n c_n$ varies as T^3, giving an overall T^4 temperature variation on the elastic constants in the low-temperature regime. However, it is not necessary to invoke elastic isotropy [131] as will now be discussed.

As discussed in Section 3.3, the lattice vibrations may be described as traveling plane waves with wave vector K. Applying periodic boundary conditions to a rectangular parallelepiped of lengths L_1, L_2, and L_3 on the sides shows that the possible

K_i values are spaced $2\pi/L_i$ apart in the i direction, *e.g.*, $K_1 = 0, \pm 2\pi/L_1, \pm 4\pi/L_1, \pm 6\pi/L_1, \ldots$ It is then easy to show that the density of states in K space, $D(K)$, is

$$D(K) = \frac{VK^2}{2\pi^2}, \tag{5.95}$$

where $D(K)dK$ is the number of K states in the range from K to $K + dK$, and $V = L_1L_2L_3$. Needed, however, is the density of states in terms of the frequency, $D(\omega)$. In the Debye approximation the frequency is taken to be linearly related to the wave vector, *i.e.*, the long-wavelength limit.

An oversimplified example will be discussed first, which will then be extended to the case at hand. For the linear dispersion relation take

$$\omega = vK \tag{5.96}$$

where v is the wave velocity. In an elastic 3D solid there are *three* waves for each distinct K value, one longitudinal (or quasi-longitudinal), and two transverse (or quasi-transverse). For the moment, the unrealistic assumption is made that all three waves have the same velocity. Then,

$$D(\omega)d\omega = 3D(K)dK \tag{5.97}$$

Combining Equations 5.95, 5.96, and 5.97 and using $d\omega = vdK$ results in

$$D(\omega) = \frac{3V\omega^2}{2\pi^2 v^3}. \tag{5.98}$$

where $D(\omega)d\omega$ is the number of modes in the range from ω to $\omega + d\omega$. This result illustrates general features, but is too simple for the present need. A first step is to recognize that the different propagation modes do not have the same velocity. Using three different relations such as Equation 5.96 gives

$$D(\omega) = \frac{V\omega^2}{2\pi^2} \sum_{p=1}^{3} \frac{1}{v_p^3} \tag{5.99}$$

where p corresponds to the three different polarizations: one quasi-longitudinal and two quasi-transverse. This result is still over simplified, because the directional dependences of the velocities have been ignored, *i.e.*, the material is still taken to be isotropc.

Another generalization achieves the desired result. The velocities, as well as Grüneisen parameter terms, depend on the direction of K in the crystal. One more density of states will be introduced; $d(\omega, \theta, \phi)$ is defined so that $d(\omega, \theta, \phi)d\Omega d\omega$ is the number of states in the range of ω to $\omega + d\omega$ with K lying is a solid angle $d\Omega$ along the direction specified by θ and ϕ. The angles are the usual ones used in spherical coordinates. Then

$$d(\omega, \theta, \phi) = \frac{1}{4\pi} \frac{V\omega^2}{2\pi^2} \sum_{p=1}^{3} \frac{1}{v_p^3(\theta, \phi)}. \tag{5.100}$$

The $1/4\pi$ is for normalization. As a check, note that if $v_p^3(\theta,\phi)$ is constant (independent of θ and ϕ) then integrating $d(\omega,\theta,\phi)$ over 4π steradians just gives Equation 5.99.

The stage is set for calculating the sums in Equation 5.94. It is assumed that the Grüneisen parameter terms depend on p, θ, and ϕ, but not on ω. Considering one of the sums,

$$\sum_n \gamma_n \left(u_n - \frac{\hbar\omega_n}{2}\right) = \int d\Omega \int d\omega \left[\frac{\tilde{V}^o\omega^2}{8\pi^3}\left(\sum_{p=1}^{3}\frac{1}{v_p^3(\theta,\phi)}\right)\gamma_{n,p}(\theta,\phi)\left(\frac{\hbar\omega}{exp\frac{\hbar\omega}{k_BT}-1}\right)\right]$$
$$= \frac{\tilde{V}^o}{8\pi^3}\int d\Omega \left(\sum_{p=1}^{3}\frac{\gamma_{n,p}(\theta,\phi)}{v_p^3(\theta,\phi)}\right)\int d\omega \frac{\hbar\omega^3}{exp\frac{\hbar\omega}{k_BT}-1}. \tag{5.101}$$

The $d\Omega$ integral is over 4π steradians ($d\Omega = \sin\theta d\theta d\phi$). The polarization and directional dependences of $\gamma_p(\theta,\phi)$ have been acknowledged with the notation. The integral over Ω will not be investigated further, because the dependences of $\gamma_p(\theta,\phi)$ on direction and polarization are unknown. (The angular dependences of the velocities are readily evaluated if the elastic constants are known.)

The evaluation of the integral over ω is discussed in Section 3.3 and in standard texts on solid state physics [39, 40]. The limits are from 0 to ω_D, the Debye frequency. By a change of variable,

$$\int_0^{\omega_D} d\omega \frac{\hbar\omega^3}{exp\frac{\hbar\omega}{k_BT}-1} = \frac{k_B^4T^4}{\hbar^3}\int_0^{x_D} dx\frac{x^3}{e^x-1}, \tag{5.102}$$

where $x_D = \theta_D/T$, and θ_D is the Debye temperature ($k_B\theta_D = \hbar\omega_D$.) As the temperature is decreased, x_D becomes large. Due to the exponential in the denominator, there is then little error in letting the upper limit go to infinity. For $T = \theta_D/10$ the error in letting the upper limit go to infinity is only about 1 percent. The integral takes the value $\pi^4/15$ in that limit. Putting everything together yields

$$\sum_n \gamma_n \left(u_n - \frac{\hbar\omega_n}{2}\right) = \frac{\tilde{V}^o\pi k_B^4T^4}{120\hbar^3}\int d\Omega\left(\sum_{p=1}^{3}\frac{\gamma_{n,p}(\theta,\phi)}{v_p^3(\theta,\phi)}\right). \tag{5.103}$$

Other terms involving $\left(u_n - \frac{\hbar\omega_n}{2}\right)$ are handled in exactly the same way. A similar procedure gives the specific heat contribution,

$$\sum_n \gamma_n^{ij}\gamma_n^{kl}c_n = \frac{4\tilde{V}^o\pi k_B^4T^3}{120\hbar^3}\int d\Omega\sum_{p=1}^{3}\frac{\gamma_{n,p}^{ij}(\theta,\phi)\gamma_{n,p}^{kl}(\theta,\phi)}{v_p^3(\theta,\phi)}. \tag{5.104}$$

Combining the various results gives

$$c_{ijkl}^T(T) = c_{ijkl}^T(0) - DT^4 \tag{5.105}$$

the well-known T^4 temperature dependence [131]. D is given by

$$D = -\frac{\pi k_B^4}{120\hbar^3}\left[\left(\frac{\tilde{c}_{ijkl}^o + \sum_r \tilde{c}_{ijklrr}^o}{\tilde{c}_{11}^o + 2\tilde{c}_{12}^o}\right)\int d\Omega\left(\sum_{p=1}^{3}\frac{\gamma_{n,p}(\theta,\phi)}{v_p^3(\theta,\phi)}\right)\right.$$
$$\left. + \int d\Omega \sum_{p=1}^{3}\frac{3\gamma_{n,p}^{ij}(\theta,\phi)\gamma_{n,p}^{kl}(\theta,\phi) + \partial\gamma_{n,p}^{ij}(\theta,\phi)/\partial\epsilon_{kl}}{v_{n,p}^3(\theta,\phi)}\right]. \tag{5.106}$$

Equations 5.105 and 5.106 are very important results. From the results of Appendix E, it is expected that the main contribution to D will be from the first term in the square brackets in Equation 5.106, and that this term is negative. Thus, D is positive and the elastic constants decrease as T^4 in the low-temperature limit.

It is seen that if all Grüneisen parameters are zero, *i.e.*, no nonlinear effects, $D = 0$ and the elastic constants do not depend on temperature. It should be remembered that the calculation is based on the assumption that the temperature is much less than the Debye temperature. A temperature $T = \theta_D/10$ results in an error of about 1% in the computation of the thermal energy, and about 3% in the specific heat, with the errors decreasing as the temperature is lowered. For a material with a low Debye temperature the T^4 range may be relatively narrow and hard to distinguish. However, for a material with a high Debye temperature, *e.g.*, diamond [151], the T^4 range may extend up to room temperature. The result is rather general, it does not assume elastic isotropy. As long as the conditions of long wavelengths and linear dispersion are met, the results could be extended to higher temperatures by integrating equations such as Equation 5.101 from 0 to x_D for various values of $x_D(= \theta_D/T)$. The integral would then depend on temperature, and the temperature dependence would no longer be T^4.

It is *possible* to obtain the correct, limiting temperature dependences by neglecting the indirect effect of the thermal expansion on the elastic constants, *i.e.*, neglecting the first term inside the square brackets in Equation 5.106, because both the first and second terms give the same limiting temperature dependence. However, such a treatment is misleading because the neglected term is the major contributor to the temperature dependence.

General Behavior

Figure 5.1 represents an effort to show the generic temperature dependence of the elastic constants for an insulating solid, with no unusual features such as a phase transition. At high temperatures the constants decrease linearly with increasing

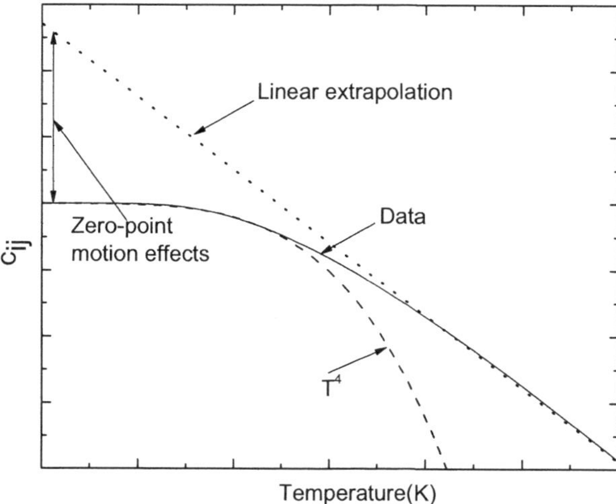

Figure 5.1 An illustration of typical regimes in the temperature dependence of elastic constants for insulating materials with no unusual features such as phase transitions. The data is well-fit at high temperatures with a linear dependence which is shown extrarpolated to 0 K. The low-temperature data is fit with the T^4 dependence. As discussed in the text, the difference between the actual $T = 0$ results and the linear extrapolation to 0 K is due to zero-point motion effects. This plot is similar to that in [131]. Note, the vertical axis just illustrates a range of the c_{ij} and does *not* start at zero.

temperature, and extrapolate to the static lattice value *with no zero-point motion contribution*, at 0 K. At low temperatures a T^4 behavior is found, which goes to the expected experimental value at 0 K, including the effects of zero-point motion.

5.6.3 Electronic Effects

Some of the key concepts regarding conduction electrons were introduced in Section 5.5.2. The internal energy due to the conduction electrons is [39, 40]

$$U_e = \int_0^\infty n(\varepsilon) f_{FD}(\varepsilon, T) \varepsilon d\varepsilon. \tag{5.107}$$

This is one of the "Fermi integrals" and rather difficult to evaluate [152]. Approximate methods are best for the present situation. At low temperatures the chemical potential, μ, approaches a constant value, the Fermi energy, ε_F, where ε_F is the highest occupied energy level at $T = 0$ [152]. At $T = 0$ all energy levels below ε_F are occupied and all above are empty. It is convenient to define a Fermi temperature as $T_F = \varepsilon_F / k_B$. Typical values of T_F for conduction electrons in metals range from 20,000 to 50,000K [39]. Thus, even at 300 K, T/T_F is of the order of 0.01 or smaller. The situation is illustrated in Figure 5.2 for $T/T_F = 1/30$.

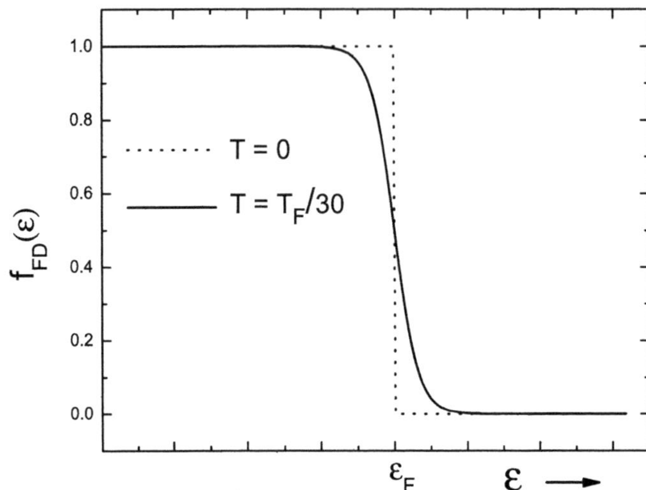

Figure 5.2 Plots of the Fermi-Dirac distribution function, Equation 5.39. At $T = 0$ all states below the Fermi energy, ε_F, are filled, all states above are empty. As the temperature is increased above $T = 0$, but for $T << T_F$, the main effect is that electrons are transferred from just below the Fermi energy to just above, as illustrated by the dotted line.

Ground State Properties

Equation 5.107 simplifies considerably for $T = 0$, because in that case $f_{FD} = 1$ for $\varepsilon < \varepsilon_F$, and 0 above ε_F. The density of states for a free-electron gas is [39]

$$n(\varepsilon) = \frac{V}{2\pi^2}\left(\frac{2m}{\hbar^2}\right)^{3/2} \varepsilon^{1/2} \tag{5.108}$$

where m is the mass of an electron and V is the volume.

The result for the free electron gas at $T = 0$ is

$$U_e^o = \int_0^{\varepsilon_F} \frac{V}{2\pi^2}\left(\frac{2m}{\hbar^2}\right)^{3/2} \varepsilon^{3/2} d\varepsilon = \frac{3}{5} N\varepsilon_F \tag{5.109}$$

with N being the number of free electrons (conduction electrons). The Fermi energy in the free-electron model is [39],

$$\varepsilon_F = \frac{\hbar^2}{2m}\left(\frac{3\pi^2 N}{V}\right)^{2/3}. \tag{5.110}$$

A basic thermodynamic relation is $P = -(\partial F/\partial V)_T$. At $T = 0$, $F = U_e^o$ and

$$P = \frac{2}{3}\frac{U_e^o}{V}. \tag{5.111}$$

The conduction electrons exert a pressure as does any gas confined to a volume V. From the definition of the bulk modulus, Equation 2.91, it follows,

$$B_e^o = \frac{5}{3}P = \frac{2}{3}\frac{N}{V}\varepsilon_F \tag{5.112}$$

where B_e^o is the contribution of the conduction electrons to the bulk modulus at $T = 0$ [40]. For many metals B_e^o is a substantial fraction of the zero-temperature, measured modulus.

Helmholtz Free Energy of the Conduction Electrons

An examination of Figure 5.2 shows that, for $T << T_F$, the temperature dependence of the electronic energy involves activity near the Fermi level; some electrons are thermally excited from states just below ε_F to states just above. Thus, it should be a good approximation to replace $n(\varepsilon)$ with $n(\varepsilon_F)$ in Equation 5.107 for calculations of the thermal energy of the electrons [153]. This is a considerable simplification which allows good approximations. Following this approach the electronic heat capacity is [39] (p. 144)

$$C_e = \frac{\partial U_e}{\partial T} = \frac{1}{3}\pi^2 n(\varepsilon_F)k_B^2 T, \tag{5.113}$$

the well-known linear temperature dependence of the electronic heat capacity. Integration of Equation 5.113 gives the electronic contribution to the internal energy,

$$U_e = U_e^o + \Delta U_e \tag{5.114}$$

with U_e^o given by Equation 5.109, and

$$\Delta U_e = \frac{1}{6}\pi^2 n(\varepsilon_F)k_B^2 T^2. \tag{5.115}$$

The objective at this point is to determine the Helmholz free energy of the electrons. For that, the entropy, S_e, is needed which is found from the basic relation between entropy and heat input,

$$S_e = \int_0^T \frac{C_e(T')}{T'}dT' = \frac{1}{3}\pi^2 n(\varepsilon_F)k_B^2 T. \tag{5.116}$$

Using $F_e = U_e - S_e T$ results in,

$$F_e = U_e^o - \frac{1}{6}\pi^2 n(\varepsilon_F)k_B^2 T^2. \tag{5.117}$$

Electronic Contribution to the Thermal Expansion

Because the isothermal elastic constants are the second derivatives of the Helmholtz free energy with respect to the appropriate strain, Equation 5.117 indicates a T^2 temperature dependence due to the electrons. However, as the discussion of Section 5.6.2 shows, the situation is a bit more subtle. Thermal expansion makes an *indirect* contribution to the elastic constant temperature dependence, and this effect is quite large. The electrons have an effect on the thermal expansion that can be important at low temperatures. The discussion will be limited to cases, *e.g.*, cubic symmetry, for which the linear thermal expansion coefficent, α_e is the same in all directions. In that case the linear expansion coefficient is 1/3 of the volume expansion coefficient [40],

$$\alpha_e = \frac{1}{3V}\left(\frac{\partial V}{\partial T}\right)_P. \tag{5.118}$$

Using basic relationships among partial derivatives,

$$\alpha_e = \frac{1}{3V}\left[-\left(\frac{\partial V}{\partial P}\right)_T\left(\frac{\partial P}{\partial T}\right)_V\right] = \frac{1}{3V}\left[\frac{-\left(\frac{\partial P}{\partial T}\right)_V}{\left(\frac{\partial P}{\partial V}\right)_T}\right]. \tag{5.119}$$

Using Equation 2.91, gives,

$$\alpha_e = \frac{1}{3B}\left(\frac{\partial P}{\partial T}\right)_V. \tag{5.120}$$

Applying $P = -(\partial F/\partial V)_T$ to Equation 5.117 and using Equation 5.109 results in

$$P = \frac{2}{3}\frac{U_e}{V}. \tag{5.121}$$

Thus, Equation 5.111 is not restricted to $T = 0$ [40]. Combining Equations 5.120 and 5.121 leads to

$$\alpha_e = \frac{2}{9BV}C_e. \tag{5.122}$$

As the temperature is increased from $T = 0$ the strain produced by the conduction electrons is given by $\xi_e = \alpha_e T$ for $\xi_e << 1$, the usual case. Using Equation 5.113 leads, finally, to

$$\xi_e = \frac{2}{27BV}\pi^2 n(\varepsilon_F)k_B^2T^2 = \frac{2}{9(c_{11}+2c_{12})V}\pi^2 n(\varepsilon_F)k_B^2T^2. \tag{5.123}$$

The last step uses Equation 2.141 for cubic symmetry. A key point here is that the thermal strain produced by the electrons is proportional to T^2.

Effects on Elastic Constants

A general discussion of the electronic effects on the elastic constants will be given in this section, without writing out all the details. The equations get a bit cumbersome, and the arguments are similar to those used in preceding sections. It should be relatively easy to fill in the details for those so interested. A first step in obtaining the electronic effects on the elastic constants is to take the second derivative of Equation 5.117,

$$\frac{\partial^2 F_e}{\partial \epsilon_m \partial \epsilon_n} = \frac{\partial^2 U_e^o}{\partial \epsilon_m \partial \epsilon_n} - \frac{1}{6}\pi^2 k_B^2 T^2 \frac{\partial^2 (\varepsilon_F)}{\partial \epsilon_m \partial \epsilon_n}. \tag{5.124}$$

The second term shows an explicit T^2 temperature dependence but, from the discussion leading to Equation 5.79, we expect an implicit temperature dependence from the first term on the right-hand side of Equation 5.124 due to thermal expansion. However, there are now two sources for the thermal expansion, the lattice vibrations and the electrons. The total strain, Equations 5.85 and 5.123, is

$$\tilde{\xi}^o_{total} = \frac{1}{\tilde{V}^o} \frac{1}{\tilde{c}^o_{11} + 2\tilde{c}^o_{12}} \left[\sum_n \gamma_n u_n + \frac{2}{9}\left(\pi^2 n(\varepsilon_F) k_B^2 T^2\right) \right]. \tag{5.125}$$

The first term in the large brackets gives a T^4 dependence for $T << \theta_D$ (Debye temperature). Thus, Equation 5.124 gives both a T^2 and a T^4 temperature dependence in the low-temperature limit. In the free-electron approximation the T^2 behavior only requires $T << T_F$, a much less stringent limitation than $T << T_D$.

It should be mentioned that the results in Section 5.6.2 need to be modified in two ways for a metal. The vibrational effects remain, but the added effects of conduction electrons must be taken into account. The thermal strain $\tilde{\xi}^o$ of Equation 5.85 should be replaced by $\tilde{\xi}_{total}$ of Equation 5.125. This substitution continues, of course, for the equations following Equation 5.85. In addition, the effects given by Equation 5.124 should be added to those of Section 5.6.2.

There is an alternate, compact way to modify the results of Section 5.6.2 to incorporate the electronic effects: 1. Make the replacement $\phi \rightarrow \phi + U_e^o$; 2. Make the replacement $\tilde{\xi}^o \rightarrow \tilde{\xi}^o_{total}$; 3. Add the term $-\frac{1}{6}\pi^2 k_B^2 T^2 \frac{\partial^2(\varepsilon_F)}{\partial \epsilon_m \partial \epsilon_n}$ to Equation 5.105.

Including electronic effects, Equation 5.105 is modified to become

$$c^T_{ijkl}(T) = c^T_{ijkl}(0) - ET^2 - DT^4 \tag{5.126}$$

where E is a constant representing the strength of the electronic contribution. A T^2 dependence of elastic constants is a general feature of conduction electrons. However, the magnitude, and even the algebraic sign, depend on features of the Fermi surface [153]. The magnitude of the effect appears to be very small in many metals, but has been observed in transition metals [154].

5.6.4 Varshni Approximation

The general behavior of elastic constants as a function of temperature, illustrated in Figure 5.1, may be approximated by a number of mathematical formulae. Varshni [155] introduced a simple equation that is particuraly good at fitting the temperature dependence of the $c_{\alpha\beta}$ for many materials. Varshni's result is obtained immediately from Equation 5.94 with a couple of approximations. First, assume that the last bracketed term on the right-hand side of Equation 5.94 may be neglected. The consideratons of Appendix E show that this is a very good approximation in many cases. Second, and more drastically, assume that the thermal vibrations can be adequately represented by just one frequency, ω_E, called an Einstein frequency after the similarity to the Einstein theory of the specific heat of solids. Using $\sum_n \gamma_n = 3N\gamma_E$ the general equation, 5.94, immediately reduces to

$$c^T_{ijkl}(T) = c^T_{ijkl}(0) - \frac{s}{\exp\frac{t}{T} - 1} \tag{5.127}$$

where $t = \hbar\omega_E/k_B$ and

$$s = -3N\gamma_E\hbar\omega_E \left[\frac{\tilde{c}^o_{ijkl} + \sum_r \tilde{c}^o_{ijklrr}}{\tilde{V}^o\left(\tilde{c}^0_{11} + 2\tilde{c}^0_{12}\right)} \right]. \tag{5.128}$$

Equation 5.127 is commonly called the Varshni equation [156]. As demonstrated in Appendix E the bracketed term in Equation 5.128 is normally negative, thus s is positive. The parameter s will, in general, be different for different elastic constants. The Varshni equation produces the correct linear dependence on temperature at high temperature. At low temperature it has an exponential temperature dependence, not the correct T^4 behavior. However, the T^4 range is in many cases rather narrow, and the difference between the Varshni Equation and the T^4 dependence can be made quite small with suitable fitting parameters. The Varshni equation usually does a good job fitting the transition region between the T^4 and T behaviors. The equation is particularly useful in cases where one wishes to subtract a "background" temperature dependence from some more complex phenomenon, such as a phase transition. The equation is of practical use, in that the elastic constants can, in many cases, be described by just three parameters over extended temperature ranges.

5.7 Phase Transitions

5.7.1 Introduction

In general, any material can exist in various states, or phases, which are distinct states due to the qualitative difference of one or more physical properties. The most well-known example is water, which can exist as a solid, liquid, or gas. The change from one phase to another is known as a phase transition. The phase transition can be driven by an externally applied parameter such as temperature, pressure, magnetic field, or electric field. There are many, many known cases, among them magnetic transitions, structural transitions, metal-insulator transitions, etc. Structural transitions often are associated with the appearance (or disappearance) of new physical properties, *e.g.*, piezoelectricity, ferroelectricity, etc.

Phase transitions can be classified as first-order (abrupt) and second-order (continuous) [58]. The terms abrupt and continuous give the more accurate description, but the terminology of first-order and second-order is firmly entrenched in the scientific language and will be used in the present discussion. At a first-order transition there is an abrupt change in the free energies, and often other physical parameters such as density, magnetization, polarization, etc. At a second-order transition these parameters change continuously as the the system passes from one phase to another. If the transition is observed as a function of temperature, then the temperature at which the transition occurs for the second-order type is called the critical temperature. Water presents a familiar example of first-order transitions. As the temperature is increased from below the melting point to above the vaporization point, there are discontinuous changes in the density at the melting temperature and at the vaporization temperature. In contrast, as the temperature of a magnetic material is increased in zero applied magnetic field, the magnetization decreases smoothly to zero at the critical temperature. (By a suitable manipulation of parameters it is possible to observe a second-order transition in water and a first-order transition in the magnetic material, but that discussion would present too great a digression at this point.)

Phase transitions are of great interest for both technological and theoretical reasons. On the technological side the historical and current importance of steam power resulting from the liquid-vapor transition in water can hardly be exaggerated. At present, the latent heat associated with the solid-liquid phase transition in a number of materials is being explored for energy storage applications. The ability of certain materials to transform to a magnetic state has had enormous implications for magnetic memory storage, magnetic devices, etc. In addition, piezoelectricity, and ferroelasticity associated with structural phase transitions have an incredible number of practical applications. Finally, phase transitions in multiferroic materials seem likely to enable entirely new classes of devices.

Phase transitions are of great theoretical interest, especially second-order transitions. Near the critical temperature very different systems behave in a very similar manner, which means that the behavior must not depend on the detailed nature of the interatomic interactions. For example the liquid-gas transition and the magnetic transition near their respective critical points look very similar when plotted against the appropriate variables [58]. Universality is the term used to describe this behavior.

The following discussion is based on the popular Landau theory of phase transitions [141]. The Landau theory presupposes a group-subgroup relation between the two phases. However, it should be remembered that there is a very large group of transitions, called reconstructive phase transitions, for which the group-subgroup relation does not hold, and thus the simple Landau theory does not apply. More sophisticated methods are required [157].

5.7.2 Landau Description of Phase Transitions

Second-Order Transitions

The Landau theory of phase transitions was originally developed [141] for second-order transitions near the critical temperature, but has since found wider use. The Landau theory focuses on a change of symmetry at the transition. One phase has higher symmetry than the other. The higher symmetry phase has all the symmetry elements (rotations, reflections, inversions, etc.) of the lower symmetry phase, plus an additional element or elements. The transiton can be described as one of symmetry breaking. The description of the transition is facilitated by the introduction of an order parameter, Q, where Q has non-zero values in the lower symmetry phase, and is usually defined so that it is zero in the higher symmetry phase. The lower symmetry phase is usually, but not always, the one occurring at lower temperature. The order parameter is defined so that it goes smoothly to zero as the critical temperature is approached in the lower symmetry phase, and remains zero in the higher symmetry phase.

An example is illustrated in Figure 5.3 [158]. It will be assumed for this discussion that the higher symmetry phase occurs at the higher temperature, which is usually true. As the temperature is lowered through the critical temperature the symmetry is lowered by the loss of several symmetry operations. The most obvious is the loss of the mirror plane perpendicular to the z axis (c axis), σ_z, but there are others. The point symmetry is lowered from 4/mmm to 4mm. There are sixteen symmetry operations associated with the 4/mmm point group, but just eight symmetry operations for the 4mm point group. The 4mm point group is a subgroup of the 4/mmm point group; all the symmetry elements of 4mm are contained in the 4/mmm point group. The Landau theory is based on this group-subgroup relationship.

As already mentioned, it is convenient to define an order parameter that differentiates the two phases, which is zero in the higher symmetry phase, but becomes non-zero in the lower symmetry phase. In the present case the distance d that the center atom moves would be a suitable order parameter. d could be considered a vector, or in the present case the z component of a vector.

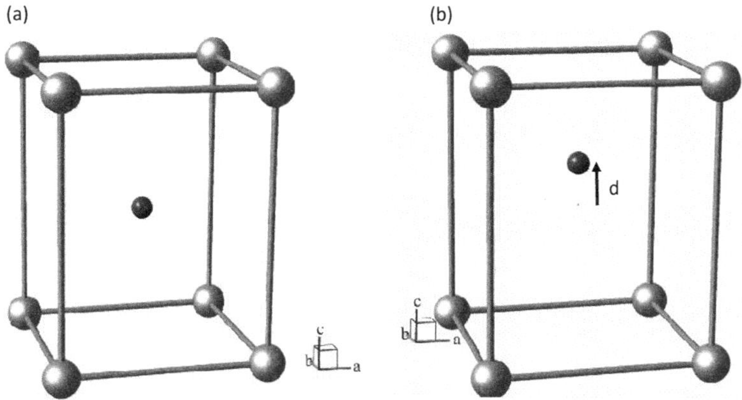

Figure 5.3 Illustration of a continuous phase transition [158]. (a) The higher symmetry phase. The tetragonal unit cell has an atom at the center of the unit cell. The point symmetry is 4/mmm. (b) The lower symmetry phase. The central atom has moved away from the center in the vertical direction. The symmetry is reduced to 4mm. The displacement, d, of the central atom from the center of the unit cell is indicated by the arrow. As the system passes through the critical temperature from the higher symmetry phase to the lower symmetry phase the displacement increases smoothly from zero. The displacement is zero at the critical temperature and in the higher symmetry phase.

A more interesting case involves net charges on the atoms. If the central atom has one unit of + charge, and the eight atoms at the vertices each have one unit of − charge, then the unit cell has an electric dipole moment in the lower temperature phase. Thus, the electric polarization would also be a suitable order parameter. As the temperature is raised through the critical temperature the polarization goes smoothly to zero at the critical temperature and remains zero in the higher symmetry phase.

The continuous behavior of the order parameter near the critical temperature in second-order phase transitions leads to the expansion of the free energy in a power series in Q, the order parameter,

$$F(Q,T) = F_o(T) + a_1 Q + \frac{1}{2}a_2 Q^2 + \frac{1}{3}a_3 Q^3 + \frac{1}{4}a_4 Q^4 + \cdots \tag{5.129}$$

Equilibrium values of Q are obtained by requiring $\partial F/\partial Q = 0$, thus there can be no linear term in Q for a minimum at $Q = 0$; $a_1 = 0$. Equation 5.129 is to be valid in both phases of the material. If the higher symmetry phase contains the inversion operation, which will be assumed here, then $F(Q,T) = F(-Q,T)$ for this phase, but this can only be true if the coefficients of the odd powers of Q are zero. Thus, in addition to a_1, $a_3 = a_5 ... = 0$. To describe the transition, it is desired that the minimum of $F(Q,T)$ with respect to Q be at $Q = 0$ in the higher symmetry phase,

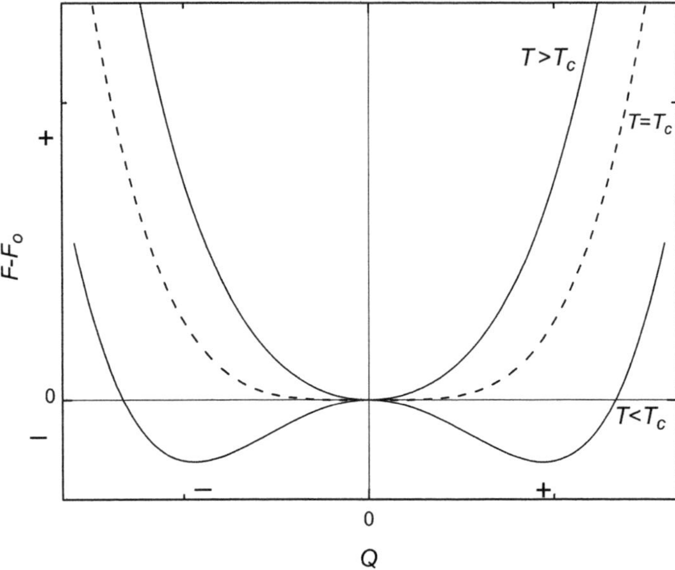

Figure 5.4 Behavior of the Landau free energy for a second-order (continuous) phase transition. Equation 5.130 is illustrated for the case $a_4 > 0$. For $T > T_c$ there is a single minimum located at $Q = 0$. As T is lowered below T_c, two minima develop on both sides of $Q = 0$ and move continuously away from the origin as the temperature is lowered. At the transition point the curve has a rather flat bottom, allowing for fluctuations at little cost in energy.

but at $Q \neq 0$ in the low temperature phase. This is accomplished if the coefficient a_2 changes sign at the transition. (All the coefficients may depend on temperature, perhaps due to thermal expansion.) This sign change is accomplished by setting $a_2 = \alpha(T - T_c)$. Then the Landau free energy expression takes the form

$$F(Q,T) = F_o(T) + \frac{1}{2}\alpha(T - T_c)Q^2 + \frac{1}{4}a_4Q^4. \tag{5.130}$$

It is assumed that coefficients other than a_2 have only a weak temperature dependence near the transition. Equation 5.130 is not expected to be valid over a very large temperature range, and certainly not down to $T = 0$.

If $a_4 > 0$, Equation 5.130 describes a second-order phase transition, as will now be discussed. Figure 5.4 illustrates the situation, which shows plots of the free energy for three different temperatures. For $T > T_c$ the plot shows one minimum located at $Q = 0$. For $T < T_c$ there are two minima located at

$$Q = \pm\sqrt{\frac{\alpha(T_c - T)}{a_4}}. \tag{5.131}$$

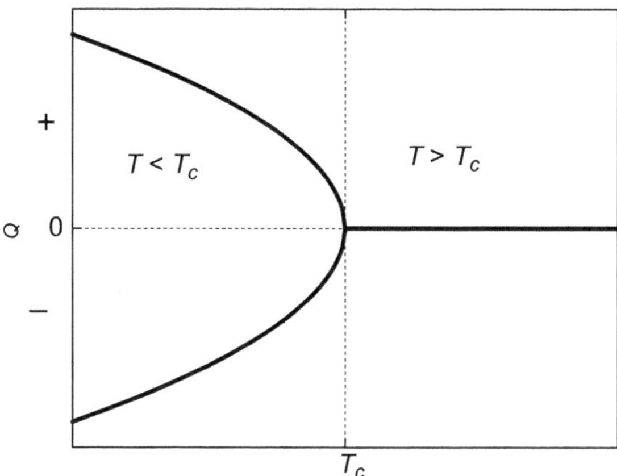

Figure 5.5 The equilibrium values of Q, Equation 5.131 and 0, *versus* temperature. Below T_c the equilibrium values move smoothly away from $Q = 0$.

These minima give the equilibrium values of the order parameter in the lower symmetry phase. Figure 5.5 illustrates the temperature dependence of the minima. As the temperature passes through T_c the minima move smoothly away from $Q = 0$. In reference to Figure 5.3, the minus values correspond to the central atom moving *downward*, in the negative z direction. Thus, the order parameter varies as $\pm\sqrt{T - T_c}$ in the low symmetry phase. This temperature dependence gives the qualitative behavior observed in many systems, *e.g.*, it describes qualitatively the behavior of the magnetization in the Ising model of magnetism, however, the actual exponent is not $1/2$. A more sophisticated treatment is needed to get the correct *critical exponent* [58].

At $T = T_c$ the free energy plot has a rather flat bottom near $Q = 0$. Although the simple Landau theory does not explicitly treat fluctuations, the flat bottom shows that the order parameter can fluctuate with little cost in free energy. Thus, large fluctuations in the order parameter are expected, and observed, near the critical temperature.

Further exploration of the Landau theory for second-order transitions shows that there is no latent heat associated with such transitions [56]. In addition, these transitions do not show hysteresis.

First-Order Transitions

Essential features of the original Landau theory for phase transitions are not satisfied for first-order transitions where abrupt changes in the order parameter occur at the transition. Nevertheless, the Landau theory is often applied to first-order transitions as well [158], and does provide a convenient way of visualizing these phase changes. Equation 5.130 is modified for the case of first-order transitions by adding a term in Q^6 and requiring the coefficient of Q^4 to be negative.

$$F(Q,T) = F_o(T) + \frac{1}{2}\alpha(T - T_c)Q^2 - \frac{1}{4} \mid a_4 \mid Q^4 + \frac{1}{6}a_6 Q^6. \tag{5.132}$$

The observed values of Q can be found from $\frac{\partial F(Q,T)}{\partial Q} = 0$, giving $Q = 0$ and,

$$Q = \pm\sqrt{\frac{\mid a_4 \mid + \sqrt{a_4^2 - 4a_6\alpha(T - T_c)}}{2a_6}}. \tag{5.133}$$

Setting the derivative $= 0$ also yields maxima for $F(Q,T)$, but these equations are not shown. Above a temperature T_1, found by setting the term under the smaller square root $= 0$, Equation 5.133 does not give real values and the only minimum is at $Q = 0$. Carrying out the indicated procedure gives

$$T_1 = T_c + \frac{a_4^2}{4\alpha a_6}. \tag{5.134}$$

The situation is illustrated in Figure 5.6. Curve b is for $T = T_1$. At higher temperatures the only minimum is at $Q = 0$. At lower temperatures two additional minima develop at $\pm Q$ as given by Equation 5.133. These two additional minima are metastable until the temperature is lowered to T_o, defined as the temperature at which these two minima are at $F - F_o = 0$ the same as for the minimum at $Q = 0$, curve c in Figure 5.6. The temperature T_o is found by requiring $F(T_o) - F_o = 0$ and $\partial F/\partial Q = 0$, resulting in

$$T_o = T_c + \frac{3a_4^2}{16\alpha a_6}. \tag{5.135}$$

Until the temperature is lowered to T_o the absolute minimum remains at $Q = 0$ As the temperature passes through T_o the absolute minima jump to non-zero values of Q as illustrated in Figure 5.6. These new absolute minima deepen as the temperature is lowered. At $T = Tc$, curve e, the minimum at $Q = 0$ becomes a maximum. Below T_c there are just two minima.

The situation is further illustrated in Figure 5.7 giving the minima as a function of temperature. Above a temperature T_1, given by Equation 5.134, there is only one minima, which is located at $Q = 0$. Below T_1 there are two additional minima, located on both sides of the original minimum, which deepen as the temperature is lowered. At the temperature T_o, Equation 5.135, the three minima achieve the same depth. Between T_1 and T_o the minimum at $Q = 0$ is the deepest, *i.e.*, the absolute minimum. Below T_o the two minima located at $Q \neq 0$ are the absolute minima. The two absolute minima deepen as the temperature is further lowered,

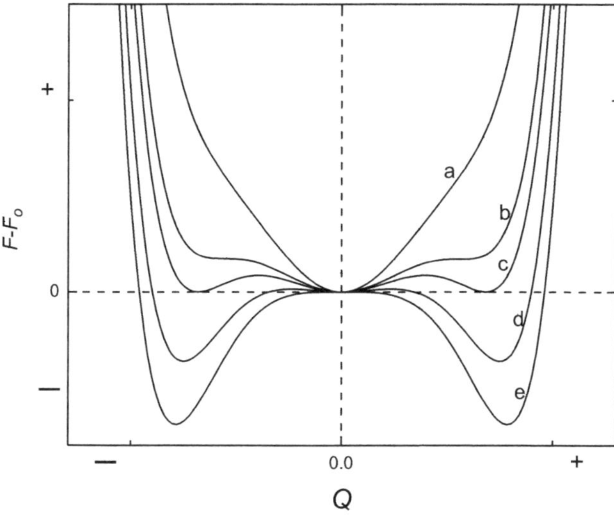

Figure 5.6 Behavior of the Landau free energy for a first-order (abrupt) transition. Equation 5.132 is illustrated for various temperatures. (a) $T > T_1$. The only minimum is at $Q = 0$. (b) $T = T_1$. Two secondary minima (for + and − Q) are developing for $Q \neq 0$. (c) The secondary minima deepen as the temperature is lowered. At a particular temperature, defined as $T = T_o$ and discussed in the text , the secondary minima have the same depth as the minimum at $Q = 0$. (d) $T < T_o$ The secondary minima have continued to deepen and now represents the stable equilibrium values of Q. (e) $T = T_c$ The secondary minima have deepened more, and the previous minimum at $Q = 0$ is now a maximum. Three minima coexist for $T_1 > T > T_c$ (from b to e in the plot). For temperatures $T < T_c$ there are two stable minima located equidistant from $Q = 0$.

and the minimum at $Q = 0$ becomes shallower. At $T = T_c$ the minimum at $Q = 0$ disappears, in fact it becomes a local maximum. Below T_c there are only two minima. In contrast to the case for second-order transitions, the changes in the order parameter for first-order transitions are abrupt.

In summary, between T_1 and T_o the absolute minimum is at $Q = 0$ and the minima given by Equation 5.133 are metastable. Below T_o the situation is reversed. The phase transition might occur anywhere between T_1 and T_c, but will occur at T_o if the system is able to transition immediately to the absolute minimum. There is likely to be considerable hysteresis.

Elastic Constants at Second-Order Transitions

Elastic constants usually show large changes at phase transitions in solids. For second-order transtions theses effects can be treated by adding two terms to Equation 5.130 : F_c for coupling between the order-parameter and the elastic

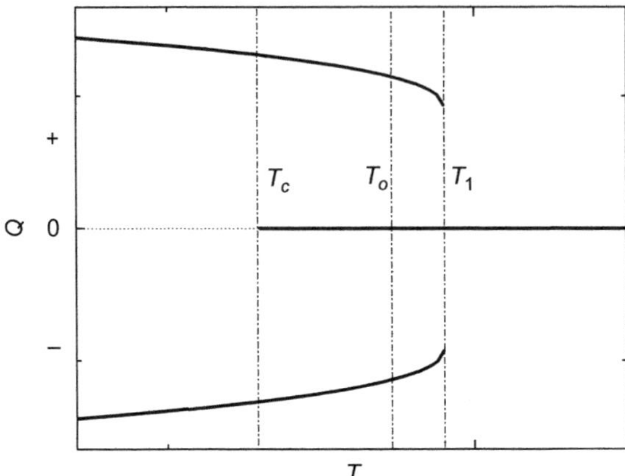

Figure 5.7 The equilibrium values of Q, Equation 5.133 and 0, *versus* temperature for a first-order transition. Above T_1 there is only one minima located at $Q = 0$. Below T_1 secondary minima develop equidistant from the $Q = 0$ minimum, but the absolute equilibrium configuration remains at $Q = 0$ until a temperature defined as T_o is reached. Below T_o the minima given by Equation 5.133 represent the absolute equilibrium configurations. Below T_c the minimum at $Q = 0$ ceases to exist, and in fact becomes a local maximum.

strains: and, F_e for the elastic energy, Equation 2.69 [159, 160]. A particular example will be explored in detail, and results will be given for other cases.

Bilinear coupling For the present example the coupling is given by

$$F_c = \beta Qe \tag{5.136}$$

where β is the coupling constant and only one strain, labeled e, will be considered for this example. With these additional terms, Equation 5.130 becomes

$$F(Q,T) = F_o(T) + \frac{1}{2}\alpha(T - T_c)Q^2 + \frac{1}{4}a_4Q^4 + \beta Qe + \frac{1}{2}c_oe^2, \tag{5.137}$$

where c_o represents the elastic constant as it would be without the phase transtion.

Due to the coupling between order parameter and strain, a change in the strain e will result in a change in Q. It will be assumed in the following that Q can change very rapidly compared to any changes in e, so that $\frac{\partial F}{\partial Q} = 0$, *i.e.*, the observed Q will be the equilibrium value [161]. The requirement of elastic stability allows another condition to be imposed: $\frac{dF}{de} = 0$.

$$\frac{\partial F}{\partial Q} = \alpha(T - T_c)Q + a_4Q^3 + \beta e = 0. \tag{5.138}$$

Applying the second condition requires a bit more care, because Q will change in response to e.

$$\frac{dF}{de} = \frac{\partial F}{\partial Q}\frac{\partial Q}{\partial e} + \frac{\partial F}{\partial e}. \tag{5.139}$$

However, the first term on the right-hand side of Equation 5.139 is zero because $\partial F/\partial Q = 0$. Then

$$\frac{\partial F}{\partial e} = \beta Q + c_o e = 0. \tag{5.140}$$

Combining Equations 5.138 and 5.140 to eliminate e results in $Q = 0$ and

$$Q^2 = \frac{\alpha}{a_4}\left(\frac{\beta^2}{\alpha c_o} - (T - T_c)\right). \tag{5.141}$$

Equation 5.141 can be put in a more convenient form by defining

$$T^* = T_c + \frac{\beta^2}{\alpha c_o}, \tag{5.142}$$

Then

$$Q^2 = \frac{\alpha}{a_4}\left(T^* - T\right). \tag{5.143}$$

Everything is in place to calculate the elastic constant. Starting with Equation 5.140,

$$\frac{\partial^2 F}{\partial e^2} = c = c_o + \beta\frac{\partial Q}{\partial e}. \tag{5.144}$$

It remains to find $\frac{\partial Q}{\partial e}$. This can be achieved by taking the derivative of Equation 5.138 with respect to e,

$$\frac{\partial}{\partial e}\left(\frac{\partial F}{\partial Q}\right) = \alpha(T - T_c)\frac{\partial Q}{\partial e} + 3a_4 Q^2\frac{\partial Q}{\partial e} + \beta = 0, \tag{5.145}$$

resulting in

$$\frac{\partial Q}{\partial e} = \frac{-\beta}{\alpha(T - T_c) + 3a_4 Q^2}. \tag{5.146}$$

Finally,

$$c = c_o - \frac{\beta^2}{\alpha(T - T_c) + 3a_4 Q^2}. \tag{5.147}$$

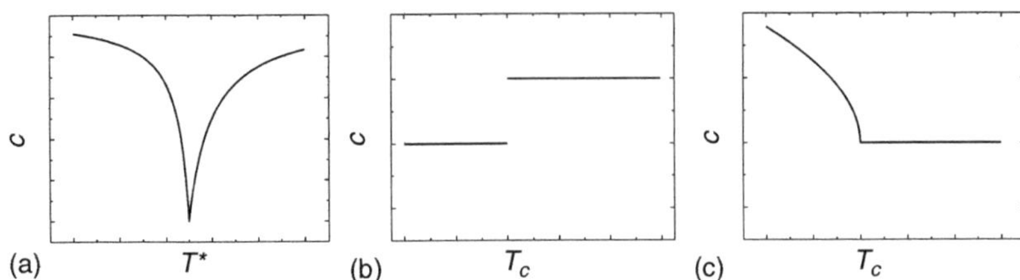

Figure 5.8 The behavior of elastic constants as a function of temperature at a second-order transition for various coupling coupling mechanisms. (a) $F_c = \beta Qe$. The minimum occurs at a temperature $T^* = T_c + \beta^2/(\alpha c_o)$. Note that the curve has the steepest slope on the low-temperature side of the transition. (b) $F_c = \gamma Q^2 e$. The step occurs at T_c. (c) $F_c = \lambda Qe^2$. The transition occurs at T_c. For this case the temperature dependence of the elastic constant, c, is the same as that of the equilibrium value of Q. Similar illustrations are found in [160].

Equation 5.142 will be used to put the results in more usual terms. It only remains to substitute the appropriate values of Q into Equation 5.147. For $T > T^*$, $Q = 0$ and, using 5.142

$$c = c_o - c_o \frac{(T^* - T_c)}{(T - T_c)}. \tag{5.148}$$

For $T < T^*$ the value of Q^2 from Equation 5.143 is used and

$$c = c_o - c_o \frac{(T^* - T_c)}{(3T^* - 2T - T_c)}. \tag{5.149}$$

For $T = T^*$ both the high-temperature and low-temperature expressions reduce to $c = 0$. In actual cases, the elastic constant is unlikely to go to zero due, *e.g.*, to higher order terms which have been neglected here, such as third-order and fourth-order elastic constants. The parameter T_c is the temperature at which the phase transition would occur in the absence of coupling between the order parameter and the strain. With the coupling in effect, the transition occurs at the higher temperature T^*. Typical behavior is illustrated in part (a) of Figure 5.8.

Coupling quadratic in the order parameter and linear in the strain In this example the coupling is

$$F_c = \gamma Q^2 e \tag{5.150}$$

where γ is the coupling constant. The derivation proceeds as in example (a), with the result

$$c = c_o - \frac{2\gamma^2}{a_4} \tag{5.151}$$

for $T < T_c$ and $c = c_o$ for $T > T_c$. There is an abrupt step downward at T_c on lowering the temperature through T_c, with no precursor effects. Similarly, there is an abrupt step upward on raising the temperature through T_c. The situation is illustrated in part (b) of Figure 5.8.

Coupling linear in the order parameter and quadratic in the strain For the third example the coupling is taken to be

$$F_c = \lambda Q e^2 \tag{5.152}$$

where λ is the coupling constant. Proceeding as before, it is found that

$$c = c_o + 2\lambda\sqrt{\frac{\alpha}{a_4}}\,(T_c - T), \tag{5.153}$$

for $T < T_c$ and $c = c_o$ for $T > T_c$. For this case, the temperature dependence of c below T_c exactly matches that of the equilibrium value of Q. Part (c) of Figure 5.8 shows typical behavior.

Elastic Constants at First-Order Transitions

As was the case in 5.7.2, the effects of the transition on the elastic constants at first-order transitions can be addressed with the Landau theory. Similar to the discussion in 5.7.2, one case will be developed, and results simply quoted for others.

Bilinear Coupling For the present case Equation 5.132 becomes

$$F(Q,T) = F_o(T) + \frac{1}{2}\alpha(T - T_c)Q^2 - \frac{1}{4} \mid a_4 \mid Q^4 + \frac{1}{6}a_6 Q^6 + \beta Q e + \frac{1}{2}c_o e^2. \tag{5.154}$$

Proceeding as before gives the solutions for Q: $Q = 0$ and,

$$Q(T)^2 = \frac{\mid a_4 \mid + \sqrt{a_4^2 - 4\alpha a_6(T - T^*)}}{2a_6} \tag{5.155}$$

where $T^* = T_c + \beta^2/(\alpha c_o)$ is the same as defined in Equation 5.142. The $Q \neq$ solution[5] holds below a temperature $T_2 = T^* + a_4^2/(4\alpha a_6)$. Still following the case for second-order transitions gives

$$c = c_o - c_o\frac{\alpha(T^* - T_c)}{\alpha(T - T_c) - 3 \mid a_4 \mid Q(T)^2 + 5a_6 Q(T)^4}. \tag{5.156}$$

Equation 5.156 comprises two solutions: one for $Q = 0$ which is valid down to T_c; and one for Q^2 given by Equation 5.155 which holds below T_2. The $Q = 0$ and

[5] As noted just below Equation 5.134 there is another $Q \neq$ solution, but that solution corresponds to a *maximum* in F.

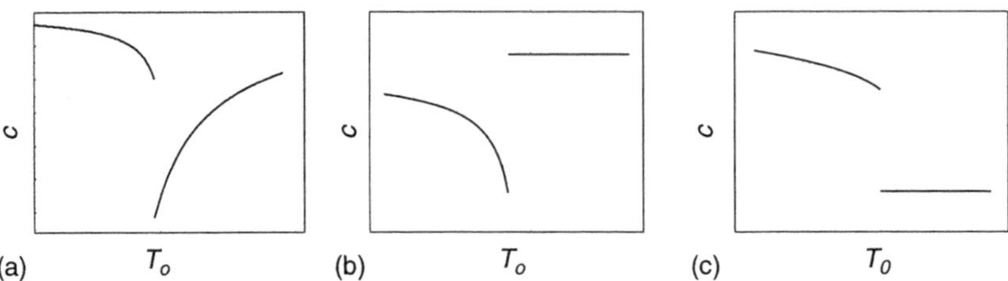

Figure 5.9 The behavior of elastic constants as a function of temperature at a first-order transition for various coupling mechanisms. The transition is indicated as occurring at a temperature T_o, which would be the case if the system always transfers to the absolute minimum. However, considerable hysteresis is possible as discussed in the text. (a) $F_c = \beta Qe$. (b) $F_c = \gamma Q^2 e$. (c) $F_c = \lambda Qe^2$. Many more cases are illustrated in [160].

$Q \neq 0$ solutions coexist between T^* and T_2. Typical behavior is shown in part (a) of Figure 5.9.

Coupling Quadratic in the Order Parameter and Linear in the Strain In this case the solutions for Q are: $Q = 0$ valid for temperatures down to T_c, and

$$Q(T)^2 = \frac{\left(| a_4 | + \frac{2\gamma^2}{c_o}\right) + \sqrt{\left(| a_4 | + \frac{2\gamma^2}{c_o}\right)^2 - 4\alpha a_6 (T - T_c)}}{2a_6} \tag{5.157}$$

valid below a temperature found by setting the quantity inside the square root sign equal to 0. The elastic constant is found to be

$$c = c_o - c_o \frac{4\frac{\gamma^2}{co} Q(T)^2}{\alpha(T - T_c) - \left(3 \mid a_4 \mid + \frac{2\gamma^2}{c_o}\right) Q(T)^2 + 5a_6 Q(T)^4} \tag{5.158}$$

Part (b) of Figure 5.9 illustrates typical behavior.

Coupling Linear in the Order Parameter and Quadratic in the Strain The solutions are

$$c = c_o \tag{5.159}$$

valid for $T > T_c$, and

$$c = c_o + 2\lambda \sqrt{\frac{| a_4 | + \sqrt{a_4^2 - 4\alpha a_6 (T - T_c)}}{2a_6}} \tag{5.160}$$

valid below a temperature found by setting the quantity inside the square root sign equal to 0. The situation is illustrated in part (c) of Figure 5.9.

5.8 Simple Quantum Systems with a Small Number of Levels

There are many cases in which a system with only a few number of levels coupled to elastic distortions have an appreciable effect on elastic constants. Examples include defects in solids and crystalline electric field effects in atoms. Relevant equations have already been given, but will be repeated here for convenience. For a system with N quantum states with energy levels E_i the partition function is, Equation 5.11

$$Z = \sum_{i=1}^{N} \exp\left(-\frac{E_i}{k_B T}\right). \tag{5.161}$$

If it is the case that there are n such independent systems per unit volume, then the Helmholtz free energy is

$$F = -nk_B T \ln(Z). \tag{5.162}$$

Finally, the effect on the elastic constants from these n entities is

$$\delta c = \frac{\partial^2 F}{\partial e^2} \tag{5.163}$$

For this discussion only a single strain e is considered and c is the corresponding elastic constant. In practice, F could be dependent on several different strains and several different elastic constants could be affected.

First, a simple two-level system will be considered [162]. The two-level system approach involving quantum tunneling between different configurations is widely used in the description of amorphous materials. However, for the present example, the simpler case with no tunneling is discussed, and the application is to defects in solids, not amorphous materials.

The energy levels will be taken as

$$E_1 = -be, \qquad E_2 = +be \tag{5.164}$$

where b is a proportionality constant. The situation illustrated in Figure 5.10 might be described by Equation 5.164. Two pairs of interstitial atoms are shown. The two energies correspond to the two possible orientations of the *pair*. These are representative of many such pairs distributed through the specimen. It is assumed that pairs are far enough apart that interactions between pairs may be neglected. An e_{12} strain, illustrated in Figure 2.8c, will affect the two pairs differently. For small strains it is reasonable to assume that the change in energy of a pair is linearly dependent on the strain. To be explicit, it is assumed that the energy of pair 1 is

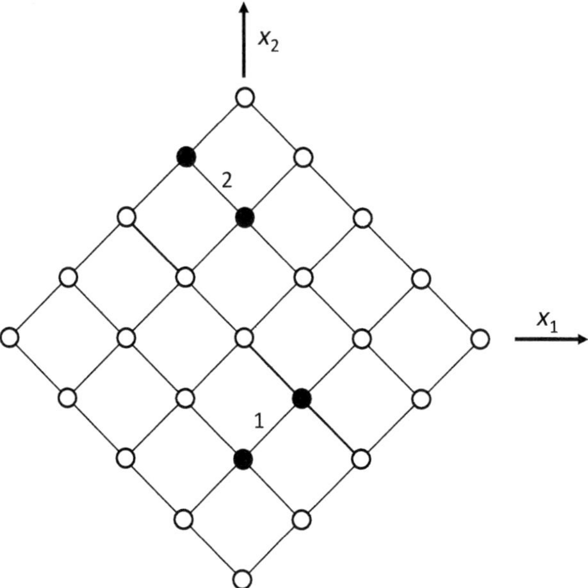

Figure 5.10 Illustration of one configuration that might be described by Equation 5.164. The open circles represent vacant interstitial sites in a lattice. (The atoms forming the lattice are not shown.) Two *pairs* of interstitial atoms are shown: 1 and 2. An e_{12} strain, illustrated in Figure 2.8c, will affect the two pairs differently, increasing the distance between the two atoms in pair 1 and decreasing that distance in pair 2.

decreased and that of pair 2 is increased, but reversing this assignment makes no difference in the final result. Applying the previous equations results in

$$\frac{\delta c}{c} = -\frac{nb^2}{k_B T c}\,\mathrm{sech}^2\left(\frac{be}{k_B T}\right) = -\frac{nb^2}{k_B T c}\,\mathrm{sech}^2\left(\frac{\Delta E}{2k_B T}\right). \tag{5.165}$$

where $\Delta E = E_2 - E_1$.

The elastic constant is decreased by the coupling of the energy levels to the strain. In the absence of strain, one would expect the number of pairs in the two configurations to be equal. However, when a strain is applied such that Equation 5.164 apply, there will be a reorientation of pairs so that there is a larger number in the lower energy state, thereby lowering the associated elastic constant. The reorientation occurs by atoms hopping among the interstitial sites, the details of which need not be of concern at this point. Implicit in the use of equilibrium statistical mechanics and thermodynamics is the assumption that the pairs are in equilibrium with the applied strain. This will be true if the period of an applied ultrasonic vibration is *much longer* than the time required for the reorientation. At low enough temperature, this assumption is unlikely to be valid; then Equation 5.165 does not apply, and $\delta c = 0$. At high enough temperatures it is likely that the pairs can reorient quickly as compared to the period of the vibration and then Equation 5.165 applies. The intermediate temperature region will be covered in Chapter 6.

The strain e in Equation 5.165 may be regarded as the equilibrium strain in the absence of the small oscillating strain due to the ultrasonic vibrations. If $e = 0$, then Equation 5.165 takes a simple form because $\text{sech}(0) = 1$. In this case the quantity $nb^2/(k_B T c)$, known as the relaxation strength, varies inversely with the temperature, the usual case. Otherwise, δc goes to zero for $k_B T << be$.

The pair-reorientation mechanism is commonly referred to as Zener relaxation [163, 90]. Figure 5.10 could represent hydrogen atom pairs occupying the octahedral sites in palladium [164, 165], except only one plane is illustrated here. An extension of the treatment is needed for the 3D case, because more than two orientations of pairs is possible.

The Snoek effect [90, 166], *e.g.*, having a C or N atom at an octahedral interstitial site in an alpha-Fe lattice, also involves reorientation effects. The distortion produced by the interstitial atom can be described as an elastic dipole [90]. The axis of the dipole can point in three different directions, and will reorient under applied stresses. The reorientation occurs by the interstitial atom hopping to a nearby interstitial site. An equation similar to Equation 5.165 applies [162] with a modification to account for more than two possible orientations of the dipole.

The preceding discussion has dealt with a two-level system, but using Equations 5.11, 5.12, and 5.29 it is possible to write down a general equation[6]

$$\delta c_{mn} = n\left[\frac{1}{Z}\left(\sum_{i=1}^{N}\frac{\partial^2 E_i}{\partial e_m \partial e_n}\exp(-E_i/k_B T) - \frac{1}{k_B T}\sum_{i=1}^{N}\frac{\partial E_i}{\partial e_m}\frac{\partial E_i}{\partial e_n}\exp(-E_i/k_B T)\right)\right.$$
$$\left. + \frac{1}{k_B T Z^2}\left(\sum_{i=1}^{N}\frac{\partial E_i}{\partial e_m}\exp(-E_i/k_B T)\sum_{i=1}^{N}\frac{\partial E_i}{\partial e_n}\exp(-E_i/k_B T)\right)\right]. \tag{5.166}$$

Equation 5.166 shows that if the energy levels and their strain derivatives are known for an entity of interest, then the effect on the elastic constants can be calculated. Equation 5.165 follows, of course, from Equation 5.166, but of greater significance, Equation 5.166 applies to more complex situations, *e.g.*, the effects on elastic constants of the crystalline electric field splitting of atomic ground states [168, 167].

[6] The last term in equation 5 of Ref. [167] should be divided by $k_B T$.

6
Ultrasonic Loss

6.1 Introduction

Ultrasonic loss (attenuation) is always present in any uses of ultrasound. For device applications, the attenuation is usually a problem to be minimized as much as possible. For materials studies, attenuation is a tool to be exploited. There are a large number of physical mechanisms that give rise to ultrasonic attenuation. Fortunately, most of these mechanisms have been covered in several excellent monographs [1, 2, 3, 90]. In view of the coverage just mentioned, the present treatment will be as follows. Several topics of general relevance to ultrasonic loss will be treated, followed by an overview of most of the known attenuation mechanisms. An overview will not be a detailed treatment, but will be an attempt to give a physical picture of the situation along with key results and references to more comprehensive coverage.

6.2 Complex Elastic Constants

Ultrasonic attenuation can be described in terms of complex elastic constants, but before discussing complex elastic constants, it seems worthwhile to discuss the use of complex quantities in general. It is common practice to use complex exponentials to describe traveling waves, because these functions are much easier to manipulate mathematically than sines and cosines. The real part is taken at the end of the calculations to compare with experimental results. Except for the Appendices and the present chapter, the space-time dependence of plane waves has been described by $e^{i(kx-\omega t)}$ for a wave traveling in the positive x direction. This choice is common in solid-state physics texts [39, 40, 41]. However, in engineering and other areas an opposite choice is frequently made, $e^{i(\omega t-kx)}$. At first, it appears that the difference between these choices is totally trivial: both expressions describe a plane wave traveling in the positive x direction; and the real parts of the two expressions equal each other. It is only when an "effect" is related to a "cause" through some response function such as a susceptibility or an elastic constant that some annoying, although not fundamental, differences arise. With proper attention to detail, physical results for the two approaches will be the same, but intermediate results may differ.

Suppose we have a complex driving "force" F^* connected to a complex "response" u^* through a complex susceptibility $\chi^* = \chi' + i\chi''$; $u^* = \chi^* F^*$. Consider the two choices:

- $$F^* = F_o e^{i(kx-\omega t)}, \tag{6.1}$$

  which gives

  $$\mathrm{Re}(u^*) = F_o\left[\chi' \cos(kx - \omega t) - \chi'' \sin(kx - \omega t)\right]; \tag{6.2}$$

  and,

- $$F^* = F_o e^{i(\omega t - kx)} = F_o e^{-i(kx-\omega t)}, \tag{6.3}$$

  which gives

  $$\mathrm{Re}(u^*) = F_o\left[\chi' \cos(kx - \omega t) + \chi'' \sin(kx - \omega t)\right]. \tag{6.4}$$

  The difference between the two expressions for the real part of u^* is that the contribution from the imaginary part of χ'' has switched signs, which amounts to a change in sign of χ''. So that the results for response functions such as complex elastic constant and complex impedance better match results in the literature, the choice of $e^{i(\omega t - kx)}$ is often made in the present chapter and in Appendix A.

The meaning of complex elastic constants will be discussed in terms of Fourier transforms of real quantities in Section 6.5. However, a discussion of many uses of this topic need not await Section 6.5. It is convenient to discuss some of these uses at this point.

One way to introduce loss into the formalism is through the use of complex elastic constants [90],

$$c^* = c' + ic'' = |c^*|e^{(i\delta)}, \tag{6.5}$$

where c^* is the complex elastic constant (subscripts are dispensed with for the present discussion). The real and imaginary parts are represented by c' and c'' respectively, $|c^*| = \sqrt{c'^2 + c''^2}$, and the phase angle $\delta = \arctan(c''/c')$.

As a simple problem, a longitudinal plane wave propagating along the x_1 axis in a high-symmetry material is considered. Combining Equations 3.3 and 6.5 gives the wave equation

$$(c' + ic'')\frac{\partial^2 u_1}{\partial x_1^2} = \rho \frac{\partial^2 u_1}{\partial t^2}. \tag{6.6}$$

A solution of the form

$$u_1 = u_1^o e^{-\alpha x} e^{i(\omega t - Kx)} \tag{6.7}$$

is substituted into Equation 6.6. The resulting equation is [11] (pages 90, 91)

$$\rho\omega^2 = (c' + ic'')(K^2 - \alpha^2 - 2i\alpha K). \tag{6.8}$$

Separating the result into real and imaginary components,

$$\begin{aligned} \rho\omega^2 &= c'(K^2 - \alpha^2) + 2c''\alpha K, \\ (K^2 - \alpha^2)c'' &- 2c'\alpha K = 0. \end{aligned} \tag{6.9}$$

Solving Equations 6.9 for α and K and assuming $\left(\frac{c''}{c'}\right)^2 << 1$ gives the approximate expressions

$$\alpha = \frac{c''K}{2c'} = \frac{c''\omega}{2\rho v^3}, \tag{6.10}$$

for the attenuation, and

$$K = \left(\frac{\omega}{v}\right) \frac{1}{\sqrt{1 + \frac{3}{4}\left(\frac{c''}{c'}\right)^2}} \tag{6.11}$$

for the dispersion relation. The imaginary part of the elastic constant has a small effect on the dispersion.

It is not difficult to show that replacing $e^{i(\omega t - kx)}$ with $e^{i(kx - \omega t)}$ in Equation 6.7 leads also to Equation 6.9 *if* the sign of c'' is switched. This result is consistent with Equations 6.2 and 6.4. The appropriate sign of c'' depends on the sign chosen in the exponential.

6.3 Measures of Ultrasonic Loss (Attenuation, Q, Internal Friction, Loss Tangent, etc.)

The ultrasonic loss can be specified in various ways. It seems worthwhile to summarize some of common methods at this point.

- For pulse echo work the attenuation coefficient α is commonly measured. A non-zero α means that the amplitude of the wave traveling in a certain direction decreases exponentially with the distance traveled. Methods for determining α from the measured echo amplitudes are discussed in Section 4.1.1. The results are usually expressed in nepers/cm, dB/cm, or dB/μ sec[1]. Equation 6.10 shows that α is directly related to the imaginary part of an elastic constant, c''.

[1] It is an easy exercise to show that if $e^{-\alpha x}$ in Equation 6.7 is replaced by $e^{-\alpha t}$, then a different expression for α is obtained which differs from that of Equation 6.10 by a factor of v.

- For resonant systems the Q is usually measured. The internal friction is defined as Q^{-1}. Equation 4.19 shows that for standing wave resonances formed by plane waves,

$$Q = \frac{\omega_n}{2\alpha v}, \tag{6.12}$$

where ω_n is a resonant frequency and v is the wave velocity. Comparing Equations 6.10 and 6.12 reveals the simple relation for the internal friction,

$$\frac{1}{Q} = \frac{c''}{c'}. \tag{6.13}$$

- Applying the complex elastic constant as given by Equation 6.5 to Hooke's Law, $\sigma = c^*\epsilon$, and assuming the strain varies as $\epsilon_o e^{-\alpha x} e^{i(\omega t - Kx)}$, the stress is given as

$$\sigma(\omega) = |c^*|\epsilon_o e^{-\alpha x} e^{i(\omega t - Kx + \delta)}. \tag{6.14}$$

The strain lags behind the stress by a time interval $\Delta t = \delta/\omega$ for positive δ (positive c'') [169]. This results illustrates that when loss is present, the stress and strain are not in phase. The loss tangent is defined as $\tan(\delta) = \frac{c''}{c''}$. Phase shifts occurring outside the specimen, due to electronics, *etc.* are *not* included in δ.

6.4 Kramers-Kronig Relations

The real and imaginary parts of c – Equation 6.5 – are not independent of each other, but are related by the Kramers-Kronig relations [9, 141, 170, 171, 172]. The following discussion and the notation follow that of Ref. [9], p. 421. Suppose $f(z)$ is a function of a complex variable[2] that is analytic on the upper half of the complex plane and on the real axis. Then the Cauchy integral formula gives,

$$f(z_o) = \frac{1}{2\pi i}\oint \frac{f(z)}{z - z_o} dz \tag{6.15}$$

if z_o lies *inside* the closed curve C. If z_o lies *outside*, then $\oint \frac{f(z)}{z - z_o} dz = 0$ by the Cauchy integral theorem.

Suppose also that

$$\lim_{|z| \to \infty} |f(z)| = 0 \tag{6.16}$$

[2] $z = x + iy$.

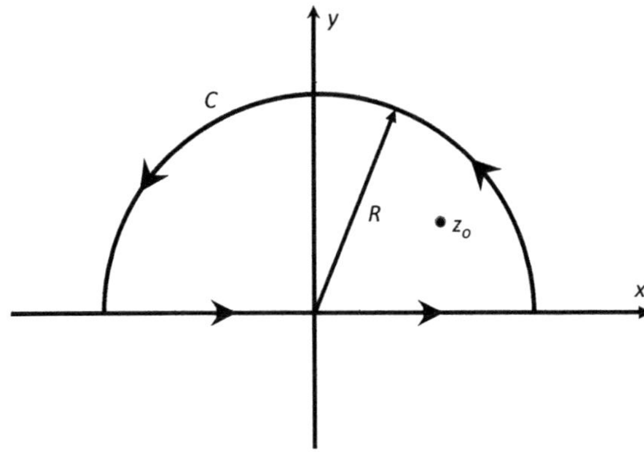

Figure 6.1 The path of integral for Equation 6.15 follows the half-circle in the upper half of the complex plane and completes the circuit along the real axis.

for z in the upper half of the complex plane. Letting $R \to \infty$, the integral over the semicircle vanishes by virtue of Equation 6.16 so that Equation 6.15 becomes

$$f(z_o) = \frac{1}{2\pi i} \int_{-\infty}^{\infty} \frac{f(x)}{x - z_o} dx. \tag{6.17}$$

Now imagine that z_o is placed on the real axis, *i.e.*, $z_o \to x_o$. The value of the integral is now taken to be the average of its value for z_o above the real axis (Equation 6.15) and its value ($= 0$) for z_o below the real axis so that,

$$f(x_o) = \frac{1}{\pi i} P \int_{-\infty}^{\infty} \frac{f(x)}{x - x_o} dx, \tag{6.18}$$

where P indicates that the principle value is to be taken due to the singularity at x_o.

Now split $f(x)$ into real and imaginary parts,

$$f(x) = u(x) + iv(x). \tag{6.19}$$

Substitution of Equation 6.19 into Equation 6.18 results in

$$u(x_o) = \frac{1}{\pi} P \int_{-\infty}^{\infty} \frac{v(x)dx}{x - x_o}, \tag{6.20}$$

$$v(x_o) = -\frac{1}{\pi} P \int_{-\infty}^{\infty} \frac{u(x)dx}{x - x_o}. \tag{6.21}$$

Next, suppose that

$$f(-x) = f^*(x), \tag{6.22}$$

giving

$$u(-x) = u(x) \qquad v(-x) = -v(x). \tag{6.23}$$

This property has several mathematical implications: the Fourier transform of $f(x)$ will be real; and, it permits changing the integration limits from $-\infty \to \infty$ to $0 \to \infty$. As x will be taken to be ω shortly, this change avoids having to deal with negative frequencies.

The following example illustrates how the limits can be changed.

$$\int_{-\infty}^{\infty} \frac{u(x)}{x - x_o} dx = \int_{-\infty}^{0} \frac{u(x)}{x - x_o} dx + \int_{0}^{\infty} \frac{u(x)}{x - x_o} dx. \tag{6.24}$$

Letting $x \to -x$ in the first integral on the right-hand side gives,

$$\int_{-\infty}^{0} \frac{u(x)}{x - x_o} dx = \int_{\infty}^{0} \frac{u(x)}{-x - x_o} (-dx) = \int_{0}^{\infty} \frac{u(x)}{-x - x_o} (dx). \tag{6.25}$$

Combining results,

$$\int_{-\infty}^{\infty} \frac{u(x)}{x - x_o} dx = \int_{0}^{\infty} \frac{u(x)dx}{-x - x_o} + \int_{0}^{\infty} \frac{u(x)dx}{x - x_o} = \int_{0}^{\infty} \frac{2x_o u(x)dx}{x^2 - x_o^2}. \tag{6.26}$$

In this way Equations 6.20 and 6.21 become,

$$u(x_o) = \frac{1}{\pi} P \int_{-\infty}^{\infty} \frac{v(x)}{x - x_o} dx = \frac{2}{\pi} P \int_{0}^{\infty} \frac{x v(x)}{x^2 - x_o^2} dx, \tag{6.27}$$

and

$$v(x_o) = -\frac{1}{\pi} P \int_{-\infty}^{\infty} \frac{u(x)}{x - x_o} dx = -\frac{2}{\pi} P \int_{0}^{\infty} \frac{x_o u(x)}{x^2 - x_o^2} dx. \tag{6.28}$$

The notation will now be changed to that appropriate to the present situation: $x_o \to \omega, x \to \omega'$, and comparing Equations 6.5 and 6.19, $u(x) \to c'(\omega'), v(x) \to c''(\omega)$. With these notational changes, the Kramers-Kronig relations become,[3]

$$c'(\omega) = \frac{2}{\pi} P \int_{0}^{\infty} \frac{\omega' c''(\omega')}{\omega'^2 - \omega^2} d\omega', \tag{6.29}$$

$$c''(\omega) = -\frac{2}{\pi} P \int_{0}^{\infty} \frac{\omega c'(\omega')}{\omega'^2 - \omega^2} d\omega'. \tag{6.30}$$

[3] There are variations on the placement of the minus sign on the right-hand side of the Kramers-Kronig relations, depending on the definition of the imaginary part of the modulus, ($c^* = c' + ic''$ or $c^* = c' - ic'$) and depending on whether the time dependence is taken to be $e^{i\omega t}$ or $e^{-i\omega t}$ [171, 172, 152]. Equations 6.29 and 6.30 are based on $c^* = c' + ic''$ and $e^{-i\omega t}$ (because $e^{-i\omega t}$ is analytic in the upper half of the complex plane). Changing c'' to $-c''$ *or* using $e^{-i\omega t}$ instead of $e^{i\omega t}$ will switch the signs in front of Equations 6.29 and 6.30. Obviously, changing both will have no effect.

Actual elastic constants usually include an approximately constant term, sometimes called c_∞. Such a term is not included in Equations 6.29 and 6.30 because it does not satisfy Equation 6.16. This problem can be handled in two ways: (i) apply Equations 6.29 and 6.30 to the frequency-dependent terms of interest and add c_∞ afterward; or, (ii) include c_∞ from the beginning, in which case the $c'(\omega')$ in Equations 6.29 and 6.30 should be replaced by $c'(\omega') - c_\infty$. Note that the Kramers-Kronig relations are nonlocal in that the effect on c'' at a particular frequency depends on the values of c' at *all* frequencies, and vice versa.

The Kramers-Kronig relations show that if there is a frequency-dependent loss, then there will be a frequency-dependent term in the real part of the elastic constant and hence a frequency-dependent sound velocity. This effect on the sound velocity is often called dispersion, and so the Kramers-Kronig relations are sometimes called dispersions relations.

The derivation presented here depends only on the properties of an analytic function of a complex variable. The derivation can also be based on causality [170].

A detailed example might be useful at this point. As shown later in Section 6.6, the complex elastic constant for Debye relaxation may be written,

$$c' = c_U - \frac{\Delta c}{1 + \omega^2\tau^2}, \tag{6.31}$$

and

$$c'' = \frac{(\Delta c)\omega\tau}{1 + \omega^2\tau^2}. \tag{6.32}$$

Equation 6.30 will be applied to the present case.

$$\begin{aligned}
c''(\omega) &= \frac{2}{\pi} P \int_0^\infty \frac{\omega}{\omega'^2 - \omega^2} \left(\frac{-\Delta c}{1 + \omega'^2\tau^2} \right) d\omega' \\
&= -\frac{2\Delta c}{\pi} \lim_{\delta \to 0} \left[\int_0^{\omega-\delta} \frac{\omega}{(\omega'^2 - \omega^2)(1 + \omega'^2\tau^2)} d\omega' \right. \\
&\quad \left. + \int_{\omega+\delta}^\infty \frac{\omega}{(\omega'^2 - \omega^2)(1 + \omega'^2\tau^2)} d\omega' \right] \\
&= -\frac{2\Delta c}{\pi} \lim_{\delta \to 0} \left\{ \left(\frac{1}{2(1 + \omega^2\tau^2)} \right) \left[\ln(\delta) - \ln(-\delta) + \ln\left(\frac{2\omega + \delta}{2\omega - \delta} \right) \right. \right. \\
&\quad \left. \left. + 2\omega\tau \left(\arctan((-\omega + \delta)\tau) + \arctan((\omega + \delta)\tau) \right) + \pi i - \pi\omega\tau \right] \right\},
\end{aligned} \tag{6.33}$$

where the limit is associated with finding the principal value of the integral.[4] Carefully taking the limit, *e.g.* $\ln(-\delta) = \ln(-1) + \ln(\delta) = i\pi + \ln(\delta)$, Equation 6.33 becomes

$$c'' = \frac{(\Delta c)\omega\tau}{1 + \omega^2\tau^2}, \tag{6.34}$$

exactly the same as Equation 6.32 which was found by quite a different method.

While giving fundamental insight into the relation between the dissipation and the dispersion of the sound velocity in acoustic systems, Equations 6.29 and 6.30 would appear to be of little practical use, because to calculate $c'(\omega)$ one needs $c''(\omega')$ at *all* frequencies, ω' – and *vice versa* – which is impractical. However, Ref. [170] presented approximate local relations, depending on the frequency at a particular point. These approximate relations have found wide usage in a number of fields.

6.5 Response Functions, Fluctuations, and Dissipation

Ultrasonic attenuation and velocity changes will be viewed from a more formal approach in this section, involving response functions, generalized susceptibilities, and the connection of these functions to fluctuations, utilizing the fluctuation-dissipation theorem. The results are quite general, but the fluctuation results are of particular use near phase transitions where fluctuations become especially important.

A very simple example will be used to illustrate the general ideas. Suppose we have a simple harmonic oscillator of a mass m on a spring, with no damping, that is struck at time zero with an impulsive force F_o of duration Δt such that $F_o\Delta t = P$, the impulse. In the absence of the force F_o the equation of motion of the oscillator is $m\ddot{x} + kx = 0$ with a solution $x = A\cos(\omega t) + B\sin(\omega t)$ where $\omega = \sqrt{k/m}$. We assume that the oscillator is initially at rest at $x = 0$. Further, we assume that Δt approaches 0 while F_o increases so that $F_o\Delta t = P$ holds with P constant. Immediately after the force is applied the mass will have a velocity $v_o = F_o\Delta t/m$, but since the distance the mass moves during the impulse is proportional to Δt^2 it is safe to assume that $x_o \approx 0$. The resulting solution satisfying the initial conditions is

$$x(t) = \frac{F_o\Delta t}{m\omega}\sin(\omega t) \qquad t > \Delta t \approx 0. \tag{6.35}$$

4 With the sign conventions used in Section 6.6 ($c^* = c' + ic''$ and $e^{i\omega t}$) the + sign, not the – sign, is appropriate [171, 172] for the right-hand sign of Equation 6.33.

It is assumed that we are dealing with a *linear* system, thus it is permissible to superimpose the solutions for a series of impulsive forces,

$$x(t) = \frac{1}{m\omega} \sum_{j}^{N} F(t_j)\Delta t \sin[\omega(t - t_j)]. \quad (6.36)$$

As $\Delta t \Rightarrow 0$ the sum may be replaced by an integral

$$x(t) = \int_0^t dt' F(t') \frac{\sin[(\omega(t - t')]}{m\omega}. \quad (6.37)$$

This equation is sometimes written

$$x(t) = \int_0^t dt' F(t') G(t - t'). \quad (6.38)$$

where $G(t - t')$ is called the Green's function. $G(t - t')$ is also called a response function because it gives the response to $x(t)$ at time t due to a impulse applied at time $t' \leq t$. It is also sometimes called a susceptibility. Note that t' cannot be greater than t, because that would imply that the response preceded the cause, *i.e.*, it would violate causality.

The objective in the following discussion is to obtain the stress response to an arbitrary strain *vs.* time profile. (The strain response to an arbitrary stress *vs.* time profile could be found in a similar manner.) The discussion *could* proceed exactly as that above where the Green's function was introduced if we imagine using strain impulses. However, as should become clear in the discussion of Section 6.6 it is more insightful in the present case to consider step functions. The basic idea is that the response to a stress, or strain, is often not instantaneous.

Suppose a step function *stress* is applied at time t'. Then, the strain at some later time t might be expressed as

$$\epsilon(t) = s(t - t')\sigma(t'). \quad (6.39)$$

The strain may continue to change for some time after the application of the stress. Such a process is called creep [90, 169]. The compliance $s(t - t')$ is taken to be zero for $t' > t$. The cause cannot precede the effect.

For the present discussion it is convenient to assume that a step function *strain* is applied at time t' and examine the resulting stress,

$$\sigma(t) = c(t - t')\epsilon(t'). \quad (6.40)$$

To be explicit, $c(t - t')$ is the time dependent elastic constant which gives the stress at time t due to a step function strain applied at time $t' \leq t$. The function $c(t - t')$ is sometimes called the relaxation modulus. As in the earlier discussion, only simple stresses and strains are considered; subscripts are omitted. The discussion is easily extended to the anisotropic case.

Now assume that a series of step strains is applied[5]

$$\sigma(t) = c(t - t_1)\epsilon_1 + c(t - t_2)\epsilon_2 + c(t - t_3)\epsilon_3 + \cdots \tag{6.41}$$

It is important to remember that $c(t - t_i)$ is zero for $t < t_i$. If the strain variation is applied in a continuous manner, the sum can be converted to an integral.[6]

$$\sigma(t) = \int_{-\infty}^{t} c(t - t')d\epsilon(t') = \int_{-\infty}^{t} c(t - t')\frac{d\epsilon(t')}{dt'}dt'. \tag{6.42}$$

It is assumed that for $(t-t')/\tau_R >> 1$, where τ_R is some characteristic relaxation time, $c(t - t')$ will reach an equilibrium or "relaxed" value, c_R. As a result, it is convenient to separate any possible c_R from the transient response [90, 169],

$$\sigma(t) = c_R\epsilon(t) + \int_{-\infty}^{t} \left[c(t - t') - c_R\right]\frac{d\epsilon(t')}{dt'}dt'. \tag{6.43}$$

Next, to determine the sinsusoidal response, set $\epsilon(t) = \epsilon_o e^{i\omega t}$. In addition, it is useful to make a change of variable, setting $\tau = t - t'$. As a result

$$\sigma(t) = \epsilon_o e^{i\omega t}\left[c_R + i\omega\int_0^{\infty} (c(\tau) - c_R)e^{-i\omega\tau}d\tau\right]. \tag{6.44}$$

Next the Euler identity for the exponential term is used,

$$\sigma(t) = \epsilon_o e^{i\omega t}\left[c_R + \omega\int_0^{\infty} (c(\tau) - c_R)\sin(\omega\tau)d\tau + i\omega\int_0^{\infty} (c(\tau) - c_R)\cos(\omega\tau)d\tau\right]. \tag{6.45}$$

Defining $c^* = \sigma(t)/\epsilon(t) = c' + ic''$ produces [90, 169, 173],

$$c'(\omega) - c_R = \omega\int_0^{\infty} (c(\tau) - c_R)\sin(\omega\tau)d\tau, \tag{6.46}$$

$$c''(\omega) = \omega\int_0^{\infty} (c(\tau) - c_R)\cos(\omega\tau)d\tau. \tag{6.47}$$

Equations 6.46 and 6.47 show the origins of the real and imaginary parts of the complex elastic constant introduced somewhat mysteriously earlier. $c^*(\omega) = c'(\omega) + ic''(\omega)$ where $c'(\omega)$ and $c''(\omega)$ are the Fourier sine and cosine transforms [9] respectively of the relaxation modulus $c(t - t')$. The relaxation modulus is a real function.

[5] It is assumed the system is linear in the sense that $\sigma(t)$ is the sum of stresses σ_1, σ_2... produced by the individual strains ϵ_1, ϵ_2... acting alone. This is known as the Boltzmann superposition principal [90].

[6] For a more careful analysis see Lakes [169] pages 19, 63.

The discussion will now turn to dissipation [58]. Let $s(x,t)$ represent an order parameter or some dynamical variable. Then,

$$s(x,t) = \int dx' \int_{-\infty}^{t} dt' \chi(x-x',t-t') f(x',t') \tag{6.48}$$

where $s(x,t)$ is the response to the excitation $f(x',t')$ and $\chi(x-x',t-t')$ is the susceptibility (Do not confuse $s(x,t)$, the response, with $s(t-t)$, the compliance, of Equation 6.39.). Ignoring any spatial dependence of χ and focusing on the time dependence

$$s(t) = V \int_{-\infty}^{t} dt' \chi(t-t') f(t') \tag{6.49}$$

where V is the volume. Set $f(t) = f_o e^{-i\omega t}$. Also, make the substitution $\tau = t - t'$. Then,

$$s(t) = f_o e^{-i\omega t} V \int_0^\infty d\tau\, \chi(\tau) e^{i\omega\tau}. \tag{6.50}$$

Now use the fluctuation dissipation theorem in the form [174],

$$\chi(t) = -\beta \frac{\partial C(t)}{\partial t} = -\beta \frac{\partial \langle \delta s(0) \delta s(t) \rangle}{\partial t} \qquad t > 0 \tag{6.51}$$

where $\delta s(t)$ represents spontaneous fluctuations of $s(t)$ about its equilibrium value, with the system in its equilibrium state. $C(t) = \langle \delta s(0) \delta s(t) \rangle$ is a correlation function. As usual $\beta = 1/k_B T$.

Then,

$$s(t) = -f_o e^{-i\omega t} V \beta \int_0^\infty d\tau \frac{\partial C(\tau)}{\partial \tau} e^{i\omega\tau}. \tag{6.52}$$

Integration by parts gives,

$$s(t) = -f_o e^{-i\omega t} V \beta \left[\left[C(\tau) e^{i\omega\tau} \right]_0^\infty - i\omega \int_0^\infty d\tau\, C(\tau) e^{i\omega\tau} \right] \tag{6.53}$$

Define the Fourier transformed susceptibility as

$$\begin{aligned} \widetilde{\chi}(\omega) &= \frac{s(t)}{f_o e^{-i\omega t}} = V \int_0^\infty d\tau\, \chi(\tau) e^{i\omega\tau} \\ &= -V\beta \left[C(\tau) e^{i\omega\tau} \right]_0^\infty + iV\beta\omega \int_0^\infty d\tau\, C(\tau) e^{i\omega\tau}. \end{aligned} \tag{6.54}$$

Define

$$\widetilde{\chi}(\omega) = \chi'(\omega) + i\chi''(\omega). \tag{6.55}$$

then [58], p. 228.

$$\chi''(\omega) = V\beta\omega \int_0^\infty d\tau C(\tau)\cos(\omega\tau) = V\frac{\beta\omega}{2}\int_{-\infty}^{\infty} d\tau C(\tau)e^{i\omega\tau}, \tag{6.56}$$

or [58]

$$\chi''(\omega) = V\frac{\beta\omega}{2}\widetilde{C}(\omega). \tag{6.57}$$

The preceding discussion has been rather general. For the present case the susceptibility is the complex elastic constant and the correlation involves the fluctuating internal stresses, $\Delta\sigma(t)$. Then $\chi''(\omega) \to c''(\omega)$ leading to

$$c''(\omega) = \frac{V\beta\omega}{2}\int_{-\infty}^{\infty} d\tau \,\langle\delta\sigma(0)\delta\sigma(\tau)\rangle\, e^{i\omega\tau}. \tag{6.58}$$

Using Equations 6.10 and 6.13 results in [159]

$$\frac{1}{Q} = \frac{V\omega}{2k_B T\rho v^2}\int_{-\infty}^{\infty} d\tau \,\langle\delta\sigma(0)\delta\sigma(\tau)\rangle\, e^{i\omega\tau} \tag{6.59}$$

and

$$\alpha = \frac{\omega^2 V}{4k_B T\rho v^3}\int_{-\infty}^{\infty} d\tau \,\langle\delta\sigma(0)\delta\sigma(\tau)\rangle\, e^{i\omega\tau}. \tag{6.60}$$

As is typical of the fluctuation-dissipation theorem, Equations 6.58, 6.59, and 6.60 are remarkable in that they connect a non-equilibrium property, dissipation (the left-hand side of the equations), to equilibrium properties, the thermal fluctuations of the stress in the equilibrium state of the system. The fluctuation theorem is quite general so that any dissipation can be associated with random fluctuations. However, Equations 6.59 and 6.60 seem to find the most use in terms of order parameter fluctuations (which are coupled to the stress) near phase transitions. Fluctuations also have an effect on the real part of the elastic constant [159, 175].

6.6 Relaxational Attenuation

Hooke's Law, Equation 2.42, implies an instantaneous relationship between stress and strain at a given point in space. However, as discussed in general in the previous section, this is often not the case. Such time dependence of the elastic response will now be discussed in more detail.

Suppose, for example, that a solid initially in equilibrium is subjected to a sudden stress. Part of the system will respond instantaneously, but there may be internal variables that require some time to reach equilibrium with the newly imposed stress. The change of an internal variable, or order parameter, toward the new equilibrium may be regarded as a stress-induced relaxation. This sort of behavior has been discussed by Zener [176] and also Nowick and Berry [90].

The term used by Nowick and Berry [90] to define the systems of present interest is anelasticity. Anelastic materials have the following properties.

- The strain (stress) response to an applied stress (strain) has a unique equilibrium value. In particular, when the applied stress or strain is released, the system completely recovers to its original state.
- The relation between stress and strain is linear.
- The response to an applied stress or strain is not instantaneous.

This definition excludes nonlinear elasticity, plasticity, and viscoelasticity [169]. However, there is a very wide range of interesting effects in solids that are included in this definition.

The general equation for anelasticity may be written [90]

$$a_o\sigma + a_1\dot{\sigma} + a_2\ddot{\sigma} + \ldots = b_o\epsilon + b_1\dot{\epsilon} + b_2\ddot{\epsilon} + \ldots \tag{6.61}$$

where the a's and b''s are constants, and the dots indicate time derivatives as usual. For the present discussion the tensorial nature of stress and strain is ignored; σ is a simple stress and ϵ is the corresponding strain.

In the following treatment one of the simplest examples illustrating anelasticity will be considered: a single internal order parameter; and, only the first-order time derivative. The starting equation is

$$\sigma = c_U\epsilon - br \tag{6.62}$$

where r is some internal parameter which will respond to the stress, b is a coupling constant, and c_U is an *unrelaxed* elastic constant. The parameter r is assumed to be zero when ϵ is zero. When a stress is applied r will change (relax) toward a new value dependent on ϵ. Taking the time derivative gives,

$$\dot{\sigma} = c_U\dot{\epsilon} - b\dot{r}. \tag{6.63}$$

The rate of change of r is assumed to have the simple form

$$\dot{r} = -\frac{1}{\tau_\epsilon}(r - \bar{r}) \tag{6.64}$$

where τ_ϵ has units of time – and is called the relaxation time – and $\bar{r}$ is an equilibrium value of r dependent on ϵ,

$$\bar{r} = \gamma\epsilon, \tag{6.65}$$

where γ is another constant.

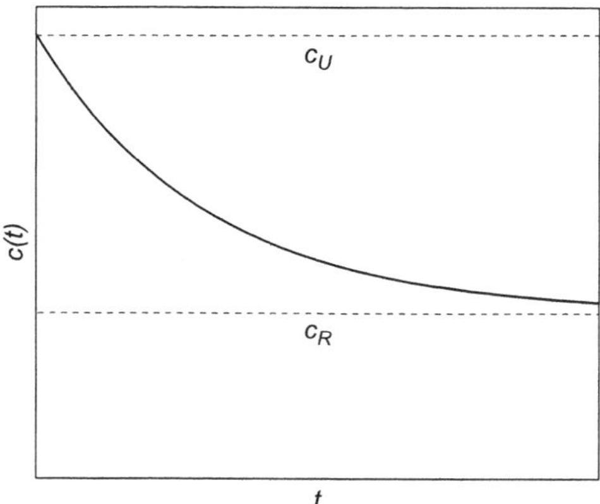

Figure 6.2 An illustration of stress relaxation. A constant strain, ϵ_o, is applied at time $t = 0$. The initial value of the stress is $c_U \epsilon_o$, but it relaxes to $c_R \epsilon_o$. The time dependent elastic constant, defined as $\sigma(t)/\epsilon_o$ is shown. The difference between c_U and c_R is exaggerated. Typical changes in the elastic constant are usually much smaller.

Combining Equations 6.62 through 6.65 results in

$$\tau_\epsilon \dot{\sigma} = c_U \tau_\epsilon \dot{\epsilon} + (c_U - b\gamma)\epsilon - \sigma. \tag{6.66}$$

For the case $\dot{\sigma} = \dot{\epsilon} = 0$, Equation 6.66 gives $\sigma = (c_U - b\gamma)\epsilon \equiv c_R \epsilon$, which defines the relaxed elastic constant c_R. The final equation is [90]

$$\sigma + \tau_\epsilon \dot{\sigma} = c_R \epsilon + c_U \tau_\epsilon \dot{\epsilon}. \tag{6.67}$$

Although the present interest in Equation 6.66 is for dynamic situations, it may be useful to examine the cases of constant stress and constant strain [177].

- A constant strain ϵ_o is applied at $t = 0$. In this case Equation 6.67 becomes

$$\sigma + \tau_\epsilon \dot{\sigma} = c_R \epsilon_o, \tag{6.68}$$

  with the solution

$$\sigma(t) = c_R \epsilon_o + (c_U \epsilon_o - c_R \epsilon_o) e^{-t/\tau_\epsilon} \tag{6.69}$$

  where $c_U \epsilon_o$ is the initial value of the stress immediately after the strain is applied. The stress relaxes exponentially toward the value $c_R \epsilon_o$ as t approaches infinity. This example shows that τ_ϵ is the relaxation time at constant strain. This example also justifies the name *relaxed* elastic constant for c_R.
- Constant stress, σ_o. Equation 6.67 becomes

$$\sigma_o = c_R \epsilon + c_U \tau_\epsilon \dot{\epsilon}, \tag{6.70}$$

with the solution

$$\epsilon(t) = \frac{\sigma_o}{c_R} - \left(\frac{\sigma_o}{c_R} - \frac{\sigma_o}{c_U}\right) e^{-t/\tau_\sigma}, \tag{6.71}$$

where $\tau_\sigma = (c_U/c_R)\tau_\epsilon$ and σ_o/c_U is the initial strain immediately after σ_o is applied. The strain exponentially approaches the value of σ_o/c_R as t approaches infinity. In materials science this effect is called creep. Obviously, τ_σ is the relaxation time at constant stress.

Now the discussion turns to dynamic effects. Assuming a time dependence of $e^{i\omega t}$, Equation 6.67 may be written,

$$\sigma(\omega) + i\omega\tau\sigma(\omega) = c_R\epsilon(\omega) + i\omega c_U\tau\epsilon(\omega), \tag{6.72}$$

which leads to

$$\sigma(\omega) = \frac{c_R + i\omega\tau c_U}{1 + i\omega\tau}\epsilon(\omega). \tag{6.73}$$

The relaxation time appearing in Equation 6.73 is τ_ϵ, but the subscripts have been dropped, as there is usually not much difference between τ_ϵ and τ_σ. The case of significant differences has been treated elsewhere [177].

Combining Equations 6.5 and 6.73 and performing some algebra gives results known as the Debye equations after P. Debye's work in 1929 on dielectric relaxation

$$c' = c_U - \frac{c_U - c_R}{1 + \omega^2\tau^2}, \tag{6.74}$$

and

$$c'' = \frac{(c_U - c_R)\omega\tau}{1 + \omega^2\tau^2}. \tag{6.75}$$

Furthermore, using Equation 6.13 results in

$$\frac{1}{Q} = \Delta_R \frac{\omega\tau}{1 + \omega^2\tau^2}, \tag{6.76}$$

where the difference between c_U and c_R has been neglected in the denominator of 6.76 and,

$$\Delta_R = \frac{c_U - c_R}{c} \tag{6.77}$$

is called the relaxation strength. In cases where the plane-wave attenuation, α, is of interest rather than the internal friction, Q^{-1}, Equation 6.12 gives

$$\alpha = \frac{1}{Q}\frac{\omega}{2v}, \tag{6.78}$$

where v is the wave velocity.

In Section 5.8 effects on the elastic constants of strain coupling to multiple level systems are calculated. These calculations are for thermodynamic equilibrium. Thus, they correspond to c_R, in effect, $c_R = c_U + \delta c$, or $\delta c = -(c_U - c_R)$. Thus, Equation 5.165, for only two levels, becomes

$$\Delta_R = \frac{nb^2}{k_B T c} \operatorname{sech}^2 \left(\frac{\Delta E}{2k_B T} \right). \tag{6.79}$$

If $\Delta E << k_B T$, the relaxation strength varies as $1/T$. This results holds for multiple levels if $\Delta E_i << k_B T$ holds for all level splittings.

Although the ideas of anelasticity were originally developed for the study of the motion of defects in solids, Equations 6.74–6.78 are much more general and apply whenever the stress couples to an internal order parameter as described by Equation 6.62, and the order parameter relaxes according to Equation 6.64. Examples include the Akheiser effect in which the ultrasonic strain shifts the frequencies of the thermal phonons, which then relax toward a new equilibrium; and, the Landau-Khalatnikov theory of phase transitions where the order parameter is one describing a phase transition. Relaxation attenuation is also important in the two-level systems characteristic of amorphous materials. These particular effects will be discussed later.

6.7 Resonance Attenuation

In contrast to a relaxation behavior, there are several situations where the ultrasonic response has a resonance character. Such effects may be conveniently described in different ways depending on the physical nature of the interactions. In some cases a quantum mechanical picture is the most direct, $\hbar\omega = \Delta E$, where $\hbar\omega$ is a quantum of energy, a phonon, absorbed from the ultrasonic wave and ΔE is an energy level splitting. Such is the case for the two-level systems of amorphous materials which have both a resonance and a relaxational response. The quantum picture is also used to describe the absorption of ultrasonic phonons by either nuclear or electron spins where ΔE is due to the Zeeman interaction of the spins with an external magnetic field.

In some cases the picture is one in which the ultrasonic frequency matches a classical resonant frequency of the material of study. Such is the case for damping due to dislocation motion, where the dislocation motion is approximately described as that of a damped, driven oscillator. The ultrasonic wave is the driver and loses energy to the vibrating dislocation line.

In other cases it seems easiest to think in terms of the ultrasonic wavelength. Such is the case for conduction electrons in metals in the presence of a magnetic field. The motion of a free electron in a direction perpendicular to the magnetic field is a circle, but is more complicated in real materials where an electron on the Fermi surface is confined to that surface. Geometric resonance occurs when the size of the orbit in the direction of propagation of the ultrasonic wave (the diameter for a circular orbit) is approximately a multiple of the ultrasonic wavelength. Resonance absorption of energy from the ultrasonic wave occcurs for these wavelengths.

6.8 Velocity-Dependent Damping

There are a number of physical situations for which the ultrasonic loss can be modeled as viscous damping. In this case, in addition to the Hooke's law restoring force, there is a force proportional to the time derivative of the strain. This situation is easily handled as a special case of Equation 6.61, which becomes

$$\sigma = c\epsilon + b\dot{\epsilon}, \tag{6.80}$$

where b is the damping constant. As before, assuming an $e^{i\omega t}$ dependence,

$$\sigma = (c + i\omega b)\epsilon. \tag{6.81}$$

Comparison with 6.5, $c^* = c' + ic''$, shows that $c'' = \omega b$ for viscous damping. Further comparison with the discussion in Section 6.2, gives [11]

$$\alpha = \frac{c''\omega}{2\rho v^3} = \frac{b\omega^2}{2\rho v^3}. \tag{6.82}$$

The attenuation is proportional to ω^2. Correspondingly, the internal friction, Q^{-1}, is proportional to ω.

6.9 Qualitative Discussion of Various Sources of Loss

The previous sections of the present chapter have treated in a general way several types of ultrasonic loss: relaxational; resonance; viscous damping; and the relation of ultrasonic loss to thermal fluctuations of stress about its equilibrium value, generally zero. The discussion will now turn to specific mechanisms, but generally in a qualitative fashion. The hope is that this treatment will be useful to students and other researchers new to the field. The treatment will be for mechanisms that *absorb* energy from the ultrasound. Elastic scattering will not be discussed.

6.9.1 Thermoelastic Effects

Consider a longitudinal ultrasonic wave propagating along a high symmetry direction in a crystal of orthorhombic or higher symmetry. At an instant in time there will be regions of compression and, displaced a half wavelength away, regions of extension.[7] As ultrasonic waves are adiabatic, or nearly so, there will be a temperature rise in the region of compression and a temperature decrease in the regions of extension. As a result, there will usually be some heat flow from high temperatures to low temperatures as the system tries to relax to a thermal equilibrium state. The heat flow results in an increase in entropy and hence dissipation. This anelastic response is described by Equations 6.74–6.76, of which Equation 6.76 is repeated here.

$$\frac{1}{Q} = \Delta_R \frac{\omega\tau}{1+\omega^2\tau^2}. \tag{6.83}$$

For the present situation it has been shown for the elastic constant c_{11} that [178, 2, 1]

$$\Delta c_R^{11} = \frac{c_{11}^S - c_{11}^T}{c_{11}^T}, \tag{6.84}$$

where c_{11}^S and c_{11}^T represent the adiabatic and isothermal elastic constants, respectively.

It remains to find the relaxation time. The required time is that needed for heat (carried by phonons) to diffuse a distance L. From the heat diffusion equation $L^2 \approx Dt$, where D is the heat diffusivity. For the case of longitudinal plane waves $L \approx \lambda$. Setting $t = \tau$ and $\lambda = \tau v$, we arrive at

$$\tau \approx \frac{D}{v^2}, \tag{6.85}$$

where v is the sound velocity. (Factors of order unity were omitted in the derivation of Equation 6.85.) Thus, the internal friction for the thermoelastic effect for longitudinal waves is given by Equation 6.83 with the relaxation strength given by 6.84 and the relaxation time by Equation 6.85. Values of τ are of order 10^{-11} second or less, thus, typical ultrasonic experiments have $\omega\tau << 1$, with the result that the thermoelastic effect tends to be obscured by other loss mechanisms. However, for the very high frequencies involved in picosecond ultrasonics, the thermoelastic mechanism can make a significant contribution [179].

The situation can be quite different for flexure experiments. In this case one side of a rod or reed may be compressed and the opposite side expanded. Then, the heat must diffuse a distance d, the thickness of the specimen. In that case [90] $\tau = d^2/\pi^2 D$ and thermoelastic damping may be observed at easily achievable frequencies [169].

The thermoelastic effect can make a contribution to loss in polycrystalline materials. Adjacent, randomly oriented, anisotropic crystallites may be compressed or expanded differently by an imposed stress. As a result heat flows between the crystallites, resulting in dissipation [169, 178].

[7] There is no effect for pure shear waves as there are no regions of compression and expansion.

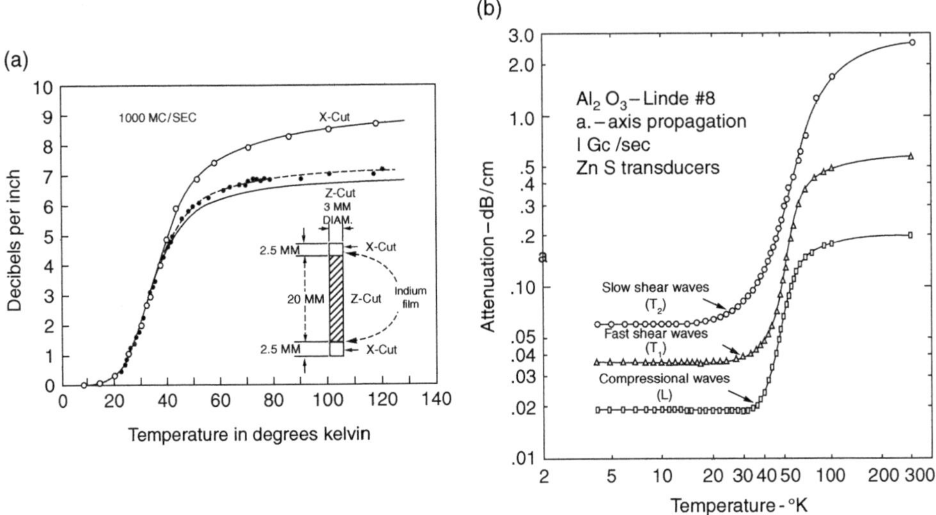

Figure 6.3 Early measurements showing attenuation due to the interactions between the ultrasonic and thermal phonons. (a) From Ref. [180] showing attenuation of 1 GHz longitudinal waves propagating along two different directions in quartz. (b) From Ref. [181] showing attenuation of both shear and compressional 1 GHz waves propagating along the *a* axis in Al_2O_3. Note the logarithmic scales.

6.9.2 Thermal Phonons

In high-quality single crystals of insulators, the dominant attenuation mechanism is expected to be due to interactions between the externally generated ultrasonic wave and thermal phonons.[8] This expectation has been borne out by numerous experiments. Figure 6.3 shows early pioneering work by Bömmel and Dransfield [180], and by de Klerk [181]. The temperature dependence of the attenuation may be roughly divided into three regimes. There is a low-temperature regime where the attenuation is very low and temperature independent. The attenuation in this regime is attributed to scattering by residual defects. Next, there is a region with a very strong temperature dependence. Finally, at high temperatures the temperature dependence is weak and the attenuation is approximately independent of temperature. The last two regions are the ones of interest, both for fundamental reasons and, also, practical considerations, *e.g.*, the intrinsic attenuation can limit the performance of micromechanical resonators [182].

Material from Section 3.2, especially Section 3.2.4, is quite relevant at this point. Several of the key points will be restated for the benefit of readers who may not have a background in solid state physics.

[8] This statement assumes there are no unusual features such as phase transitions present.

- In the harmonic approximation, Equation 3.59, the normal modes for the lattice vibrations are given by $u_\alpha(n,l,t) = A_\alpha(n)\exp[i(\vec{K}\cdot\vec{R}_l - \omega t)]$, Equation 3.64. $u_\alpha(n,l,t)$ gives the displacement from equilibrium of the atom located at (l,n). The index n refers to the nth atom in the basis. For only one atom in the basis n may be ignored. The subscript α indicates the x_i ($i = 1, 2$, or 3) coordinate.
- The allowed $\vec{K}$ values are determined by the boundary conditions and given by Equation 3.108. All unique solutions are given by $\vec{K}$ values within the first Brillouin zone. Any $\vec{K}$ vector outside the first Brillouin zone may be translated into the first zone by the addition, or subtraction, of a reciprocal lattice vector [39, 40].
- The equation connecting ω and $\vec{K}$ is called the dispersion relation. An illustration is given Figure 3.7.
- The possible energies of each normal mode are just those of a quantum harmonic oscillator,

$$E_s = (s + 1/2)\hbar\omega, \tag{6.86}$$

  where $\hbar$ is Planck's constant, ω is the angular frequency of the oscillator, and $s = 0, 1, 2, 3 \ldots$
- The thermal average energy in mode (K,p) is[9]

$$u_{K,p} = \left(\langle s_{K,p}\rangle + \frac{1}{2}\right)\hbar\omega_{K,p}, \tag{6.87}$$

  where $\langle s_{K,p}\rangle$, the thermal average number of quantum excitations in mode (K,p), is given by the Planck distribution function

$$\langle s_{K,p}\rangle = \frac{1}{\exp\left(\hbar\omega_{K,p}/k_B T\right) - 1}. \tag{6.88}$$

- It is accurate to call $\langle s_{K,p}\rangle$ the average number of quantum excitations in mode (K,p), but it is more convenient, and common practice, to speak of the number of phonons of wave vector $\vec{K}$ and polarization p in the specimen. The terminology is quite analogous to the use of photons to describe electromagnetic waves and will be used in the present section.

When anharmonic terms are included in the interatomic potential (higher order terms in Equation 3.57) the $u_\alpha(n,l,t)$ mentioned above are no longer the exact normal modes. This means that an initially pure $u_\alpha(n,l,t)$, will, after some time, become a wave that includes components of waves at other frequencies and wavelengths. In the language of phonons, this means the phonon has a finite lifetime or finite mean free path. This interaction occurs because the strain produced by one wave shifts the frequencies of other waves, and vice versa. The theoretical

[9] There are three polarizations, p, for each K value.

treatments divide naturally into two regimes: $\omega\tau_{th} >> 1$ and $\omega\tau_{th} << 1$ where ω is the angular frequency of the externally generated ultrasonic wave and τ_{th} is the lifetime of the thermal phonon. Because τ_{th} is a decreasing function of temperature, the limits $\omega\tau_{th} >> 1$ and $\omega\tau_{th} << 1$ correspond to the low-temperature and high-temperature regimes, respectively.

$\omega\tau_{th} >> 1$

The case $\omega\tau_{th} >> 1$ was first treated by Landau and Rumer [183] and later by others [184, 185]. In this case the thermal phonon lifetime is many periods of the ultrasonic wave. The wave ultrasonic frequency and wave vector may be considered as well-defined and the interaction is considered as a three-phonon process which conserves wave vector and energy ($\hbar\omega$). Consider the case where two phonons, 1 and 2, are destroyed and another, 3, is created. The inverse case is also possible. It can be shown that [184]

$$\vec{K}_1 + \vec{K}_2 = \vec{K}_3 \tag{6.89}$$

or

$$\vec{K}_1 + \vec{K}_2 = \vec{K}_3 + \vec{b}, \tag{6.90}$$

where $\vec{b}$ is a reciprocal lattice vector. The first is called an N (normal) process, and the second is called a U (umklapp) process [40]. In addition,

$$\omega_1 + \omega_2 = \omega_3. \tag{6.91}$$

Equation 6.89 or Equation 6.90 and Equation 6.91 are the crystalline counterparts to the conservation of energy and momentum for elastic collisions of free particles.

It must be remembered that $\vec{K}$ and ω are connected by the dispersion relations as shown, *e.g.*, by Figure 3.7. A study of the above conservation equations along with the dispersion curves reveals that the possible three-phonon processes are severely restricted. A clear discussion of these restrictions may be found in Ref. [2]. Considering the conservation laws of Equations 6.89–6.91, Klemens [184] finds $\alpha_t \propto \omega T^4$ for transverse waves and a very low attenuation for longitudinal waves. Although a T^4 behavior is found in a certain temperature range, overall these theoretical results are not in good agreement with the experiments [181] which show that longitudinal waves are attenuated almost as much as transverse waves. Furthermore, the temperature dependence of the attenuation is often higher than T^4. The discrepancies are largely resolved by taking into account two factors: finite values of τ_{th} and crystalline anisotropy [185]. If a phonon has a lifetime τ_{th} then its energy is uncertain by an amount $\hbar/\tau_{th}$. Thus, Equation 6.91 only has to be obeyed within limits set by the Heisenberg uncertainty principle. Crystalline anisotropy brings in other possibilities for three phonon interactions. As a result good agreement between theory and experiment is found for the ultrasonic attenuation in high-quality insulators in the $\omega\tau_{th} >> 1$ limit [185].

$\omega\tau_{th} << 1$

The case $\omega\tau_{th} << 1$ was first developed by Akhieser [186]. In this case the thermal phonon is scattered in a small fraction of a period of the ultrasonic wave. For such a short time the frequency of the ultrasonic wave is not well-defined and Equation 6.91 does not apply. Instead, the ultrasonic wave may be regarded as an essentially static strain acting on the thermal phonons. This strain will shift the frequencies of the thermal phonons. The normal equilibrium number of phonons in a mode of frequency ω is given by $(\exp(\hbar\omega/k_BT) - 1)^{-1}$, Equation 6.88. As the ultrasonic wave shifts ω, the population in a particular mode will be shifted from the equilibrium value. In general different modes will be shifted by different amounts, and phonon populations will readjust between modes with some phonon relaxation time τ_{th}. This readjustment will lag the applied stress and the result will be relaxational attenuation.

The Akheiser mechanism has been treated in detail by Woodruff and Ehrenreich [187] and Maris [185] using a Boltzmann equation approach. The results are complicated, with parameters that are difficult to calculate. Simple results are found in the limit $\omega\tau_{th} << 1$,

$$\alpha = \frac{\gamma^2\omega^2 T\kappa}{\rho v^5}, \tag{6.92}$$

where γ is a Grüneisen constant which, for longitudinal waves, is the Grüneisen constant associated with thermal expansion [187]. In general, γ^2 involves a complicated average over phonon modes [185]; however, it is a convenient adjustable parameter. Other parameters are ρ the density, v the sound velocity, and κ the thermal conductivity. For temperatures above the Debye temperature $\kappa \propto T^{-1}$ and the attenuation is independent of temperature. An approximate temperature independence is seen in the high temperature range of Figure 6.3.

6.9.3 Conduction Electrons

Conduction electrons are a significant source of loss in metals under certain conditions. It is a rather complicated topic, especially when the important case of magnetic fields is included. The effects are strongest at low temperatures. One simple example, and a brief summary of other situations will be discussed in the present section. The reader is referred to chapter 8 of Lüthi's book [3] for a comprehensive review. Features to be explained are shown in Figure 6.4. The normal state will be discussed first.

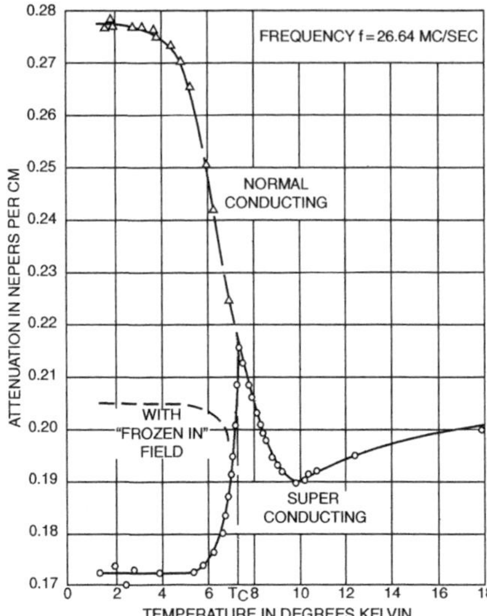

Figure 6.4 Pioneering measurements on the longitudinal-wave ultrasonic attenuation in tin [188]. Measurements on the normal state and the superconducting state are shown. The ultrasonic attenuation in the superconducting state drops drastically below the superconducting transition temperature, T_C. The normal state attenuation was observed by applying a magnetic field to suppress the superconducting transition. The normal state attenuation increases strongly as the temperature is lowered.

One of the first theoretical treatments of attenuation due to electrons was by Mason [189] who assumed the loss to be due to viscous damping by a gas of free electrons, *i.e.*, the Drude model [40]. Equation 6.82 gives the results for viscous damping in terms of a coefficient b. The viscosity can be considered as a tensor [11, 185]. We will consider the system to be elastically isotropic. We have the liberty to choose which two elastic constants are to be the independent ones. For the present case we choose c_{44} and B. With this choice, the elastic constant governing longitudinal waves may be written $c_{11} = B + 4c_{44}/3$. Taking η as the shear viscosity and χ as the volume viscosity leads to

$$\alpha = \frac{\eta\omega^2}{2\rho v^3}, \tag{6.93}$$

for the shear wave attenuation and

$$\alpha = \left(\chi + \frac{4\eta}{3}\right)\frac{\omega^2}{2\rho v^3} \tag{6.94}$$

for compressional wave attenuation. With an adjustable parameter, these equations were found to give a good approximation to the measured attenuation in lead and tin [188, 189].

Equations 6.93 and 6.94 represent a good first step toward the explanation of ultrasonic attenuation in metals. However, they are the result of a completely classical calculation and cannot represent a realistic explanation. Even free electrons are governed by Fermi statistics and the resulting Fermi sphere.

As was the case for attenuation due to thermal phonons, it is convenient to consider two different regimes. The following parameters will be important: the wave vector of the ultrasonic wave,[10] q; the ultrasonic velocity, v_s; the mean free path of an electron between collisions, l_e; the mean free time between collisions for the electron, τ; and the Fermi velocity, v_F, *i.e.*, the velocity of an electron on the Fermi surface. The two regimes are $ql_e << 1$, and $ql_e >> 1$. It is worth noting that $\omega\tau = v_s q l_e / v_F = (v_s/v_F) q l_e << ql_e$ because for typical values, $v_s/v_F \simeq 10^{-3}$.

The case of $ql_e << 1$ leads to a relaxation-type attenuation [190]. The stress field of the ultrasonic wave distorts the Fermi surface, leading to a non-equilibrium distribution. The electron distribution relaxes toward equilibrium resulting in attenuation as for the anelastic solid. For the case $ql_e >> 1$ the electron mean-free path is many times the ultrasonic wavelength. In this case electrons can "surf" along the ultrasonic wave and thereby absorb energy from the wave. Because of the great difference between the sound velocity and the Fermi velocity, the only electrons that can "surf" a considerable distance while maintaining a fixed phase relationship with the wave are those electrons whose velocities are almost perpendicular to the ultrasonic wave velocity [2].

The entire range of ql_e has been treated by Pippard [191, 192]. The results for the *amplitude* attenuation coefficient are.[11]

$$\alpha_l = \frac{nm}{2\rho v_l \tau}\left(\frac{a^2 \arctan a}{3(a - \arctan a)} - 1\right) \tag{6.95}$$

for longitudinal waves, and

$$\alpha_t = \frac{nm}{2\rho v_t \tau}\left(\frac{2a^3}{3[(1 + a^2)\arctan a - a]} - 1\right), \tag{6.96}$$

for transverse waves[12] where $a = ql_e$, n is the number of conduction electrons per unit volume, and m is the mass of an electron.

[10] It seems to be almost standard practice to use q for the ultrasonic wave vector, even if k or K is used elsewhere in the same work [1, 2, 3], when dealing with electrons. Such practice will be continued in the present work, as it seems likely to facilitate comparison of the present discussion with other work.

[11] Pippard's calculation was for "the attenuation of energy per unit length." The *amplitude* attenuation coefficient is 1/2 the energy, or power, attenuation coefficient. Because the amplitude attenuation coefficient is the one denoted by α in the present work, Pippard's equations [191] were divided by 2, thus the factor 2 in the denominators of Equations 6.95 and 6.96.

[12] Equation 6.96 appeared first in [191] in a different algebraic form. The version appearing in [192] has an obvious error. It is missing the -1 in the parenthesis.

The limits of these equations are:

$$ql_e << 1$$

$$\alpha_l = \frac{2}{15}\frac{nmq^2l_e^2}{\rho v_l \tau} = \frac{2}{15}\frac{nmv_F l_e}{\rho v_l}q^2 = \frac{2}{15}\frac{nmv_F^2}{\rho v_l^3}\omega^2\tau \tag{6.97}$$

for longitudinal wave[2], and

$$\alpha_t = \frac{1}{10}\frac{nmq^2l_e^2}{\rho v_t \tau} = \frac{1}{10}\frac{nmv_F l_e}{\rho v_t}q^2 = \frac{1}{10}\frac{nmv_F^2}{\rho v_t^3}\omega^2\tau \tag{6.98}$$

for transverse waves.

$$ql_e >> 1$$

$$\alpha_l = \frac{\pi}{12}\frac{nml_e q}{\rho v_l \tau} = \frac{\pi}{12}\frac{nmv_F}{\rho v_l}q = \frac{\pi}{12}\frac{nmv_F}{\rho v_l^2}\omega \tag{6.99}$$

for longitudinal waves[1, 2] and

$$\alpha_t = \frac{2}{3\pi}\frac{nml_e q}{\rho v_t \tau} = \frac{2}{3\pi}\frac{nmv_F}{\rho v_t^2}\omega, \tag{6.100}$$

for transverse waves. As discussed in Footnote 11, these equations are for the *amplitude* attenuation coefficient.

For most metals and measurement frequencies, room temperature is well within the regime $ql_e << 1$, *i.e.*, short electron mean free path. In that case $\omega\tau$ is exceptionally small (recall $\omega\tau << ql_e$) and the attenuation is usually completely negligible compared to other attenuation mechanism always present at room temperature. The electronic contribution to the attenuation in metals becomes appreciable, in fact usually dominant, at low temperatures where $ql_e >> 1$, Figure 6.4. The highest values of ql_e are normally found at low temperatures in materials with the fewest defects or impurities; thus, such conditions give the highest attenuation.

6.9.4 Superconductivity

One of the first experiments to show the effect of superconductivity on ultrasonic attenuation is illustrated in Figure 6.4. The ultrasonic attenuation in the superconducting state decreases rapidly below the transition. (Application of a suitably strong magnetic field suppresses the transition to the superconducting state.) In the context of the BCS theory [193], not yet published at the time of the experiments, this effect is understood as the formation of Cooper pairs below the transition, the number of pairs increasing with decreasing temperature. The Cooper pairs carry the superconducting current, but are not available for the ultrasonic attenuation. In fact, an energy gap in the electron energy spectrum forms at the transition. The gap is temperature dependent and goes to zero at the transition. The electrons in the ground state lie below the gap and are coupled in Cooper pairs. The normal electrons, the ones responsible for the attenuation, lie above the gap.

The BCS theory for the temperature dependence of the longitudinal-wave ultrasonic attenuation is

$$\frac{\alpha_s}{\alpha_n} = \frac{2}{\exp(\Delta/(k_B T) + 1} \tag{6.101}$$

where 2Δ is the temperature-dependent energy gap. The subscripts s and n correspond to the superconducting and normal states respectively. The temperature dependence of the attenuation below the gap provides a good measure of the gap. Experiments such as these provided strong evidence for the BCS theory of superconductivity, and also proved that the low-temperature attenuation in normal-state metals is, in fact, due to the conduction electrons. Further information about the use of ultrasound to study superconductivity is found in a number of good reviews [1, 2, 3, 194, 195, 196].

6.9.5 Magnetoelastic Effects

The usual theory of conduction electrons treats the electrons as plane waves. Just as was the case for phonons in Section 3.2.4, the possible wave vectors of the electron plane waves are restricted. A 2D representation of these states is given in Figure 6.5, which is similar to Figure 3.8 for phonons. A central concept in the theory of metals is that of the Fermi surface. This is a surface in $\vec{k}$ space. In 2D, the "surface"is just a circle for free electrons as shown in Figure 6.5. Three-dimensional representations can be found in most solid-state texts [40, 39]. Electrons are Fermions and thus only one electron can occupy a particular quantum state. (Two electrons can occupy a particular k state, one for spin up and one for spin down.) At zero of temperature all states are occupied up to a highest value which lies on the Fermi surface. The Fermi surface is a surface of constant energy. As the temperature is raised, electrons near the Fermi surface are thermally promoted to higher energy states. However, the thermal energy available is $\approx k_B T$, which is ≈ 0.026 eV at room temperature. Typical Fermi energies [39] are several eV, thus, even at room temperature the only electrons thermally promoted to higher energies lie in a very narrow energy band at the Fermi surface. This means that for most physical effects, the electrons involved lie very close to the Fermi surface, which demonstrates the importance of this surface.

This section will focus on the ultrasonic loss associated with the application of a magnetic field to a non-magnetic metal. The motion of electrons in the presence of

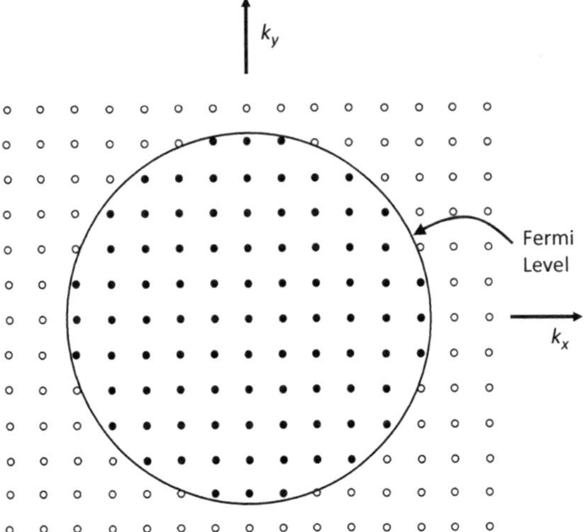

Figure 6.5 Illustration of allowed $\vec{k}$ values in 2D for free electrons. The illustration is for $T = 0$ in which case all states below the Fermi surface are filled and all states above are empty, as illustrated by the filled and empty circles, respectively.

an external magnetic field $\vec{B}$ can be described by a semiclassical model [40]. For a uniform field, the real-space motion in a plane perpendicular to the field is found to be, equation 12.35 of Ref. [40],

$$\vec{r}_\perp(t) - \vec{r}_\perp(0) = -\frac{\hbar}{eB}\hat{B} \times (\vec{k}(t) - \vec{k}(0)), \tag{6.102}$$

where $\vec{r}_\perp(t)$ is the component of the position vector in the plane perpendicular to the magnetic field, $\hat{B}$ is a unit vector in the direction of the magnetic field, and e is the magnitude of the charge on the electron. Again, from Ref. [40], the real-space orbit in a plane perpendicular to the magnetic field is just the k-space orbit rotated 90° and scaled by the factor $\hbar/eB$.

The magnetic field does no work on the electron so an electron on the Fermi surface remains on the (constant energy) Fermi surface. For free electrons the orbit will be a circle of radius $\hbar k_F/eB$ where k_F is the wave vector of an electron on the Fermi surface.

Ultrasonic measurements have been a valuable tool for exploring the Fermi surface [192, 197, 198]. At least three types of magnetoacoustic phenomena have been observed [199].

Geometric Resonance

This effect involves matching the size of the orbit of the electron with an integral or half-integral multiple of the wavelength of the sound, and is, therefore, known as a geometric resonance. The situation is illustrated in Figure 6.6. The major

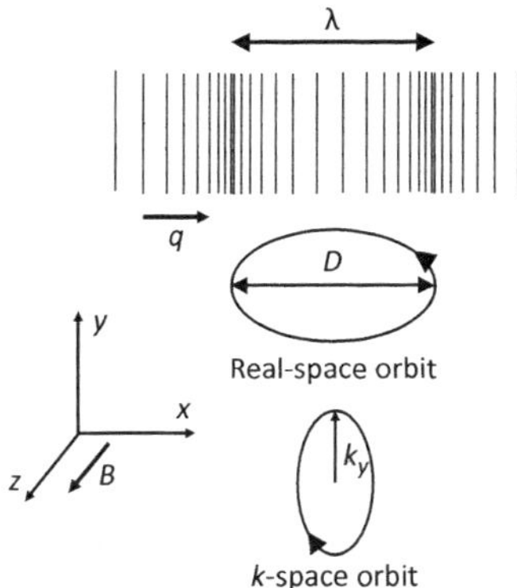

Figure 6.6 Illustration of a condition for maximum attenuation for geometric resonance. A compressional wave with wave vector q (wavelength λ) travels in the x direction. A real-space orbit and the corresponding k-space orbit are depicted. The magnetic field, B, points in the z direction, perpendicular to the plane of the orbits. D is the "size" of the noncircular real-space orbit in the x direction, the direction of q; k_y is a measure of the k-space orbit. The orbits lie in the $x-y$ plane.

contribution to the attenuation is expected to occur at the ends of the orbits. Here, the paths of the electrons are nearly parallel to the ultrasonic wavefront. Because the Fermi velocity is $>>$ than the ultrasonic velocity, the electrons see an almost stationary wavefront in this region and it is expected that the maximum absorption of energy by the electrons will occur, and hence the maximum in ultrasonic attenuation [2]. This simple analysis gives the condition for a maximum in the attenuation as $n\lambda = D$, where n is an integer. More careful analyses support this result for longitudinal waves [200, 201] and transverse waves [201]. A different analysis [3], however, gives the condition for an attenuation maximum for transverse waves as $(n + 1/2)\lambda = D$. This difference turns out not to be especially important as will be shown in what follows.

The experiments are performed by varying the magnetic field at a fixed ultrasonic frequency. Application of Equation 6.102 to the the situation of Figure 6.6 yields

$$D = \frac{2\hbar k_y}{eB}. \tag{6.103}$$

Combining either $n\lambda = D$ or $(n + 1/2)\lambda = D$ with Equation 6.103 shows that the attenuation is periodic in $1/B$ with the period

$$\Delta\left(\frac{1}{B}\right) = \frac{e\lambda}{2\hbar k_y}. \tag{6.104}$$

For a spherical Fermi surface $k_y = k_F$, but that situation is not of much interest because there are easier ways to determine k_F. For more complicated Fermi surfaces it can be seen that the attenuation of ultrasonic waves propagating along judiciously chosen directions can yield important information about the dimensions of the Fermi surface.

Passed over in the preceding discussion is that the electron orbits that dominate the attenuation are extremal orbits, *i.e.*, those for which the measured D is a local minimum or maximum [40]. Other conditions are important. For resonances to be possible it is necessary that $\lambda \leq D$ (or $D/2$). This means that the resonances will disappear at high magnetic field. Also needed are: $\omega_c\tau > 1$, where ω_c is the cyclotron frequency and τ is the average time for an electron to scatter, so that the electron will complete at least a large fraction of an orbit before scattering; and, $ql_e > 1$.

Quantum Oscillations

There are magnetic-field dependent oscillations in the ultrasonic attenuation associated with the quantization of the orbital motion of the electron. These oscillations will be summarized next. In the absence of a magnetic field, the energies of a free electron in a metal are given by

$$E = \frac{\hbar^2}{2m}(k_x^2 + k_y^2 + k_z^2), \tag{6.105}$$

where m is the electron mass. (Some non-free electron effects can be partially accounted for by replacing m by m^*, an effective mass [39, 40].) In the presence of a uniform magnetic field, taken to be in the z direction, the orbital motion becomes quantized with the result [39, 40, 202, 203]

$$E_n = \left(n + \frac{1}{2}\right)\hbar\omega_c + \frac{\hbar^2 k_z^2}{2m}, \tag{6.106}$$

where ω_c is the cyclotron frequency, which for free electrons $= \frac{eB}{m}$, and $n = 0, 1, 2, .$ Equation 6.106 is the quantum mechanical analog to the classical case. In the classical case a free particle follows a circular orbit in the plane perpendicular to the magnetic field while the motion in the z direction is unaffected by the magnetic field. Equation 6.106 describes similar motion except that the orbital motion is quantized. The quantized energy levels are usually called Landau levels.

Figure 6.7 illustrates the situation in the $x-y$ plane, for $k_z = 0$. (It is important to keep in mind that the discussion is for a cross-section of the real 3D case. The true 2D case has interesting features not found in 3D.) The original states in the absence of a field have moved to the nearest Landau level. As the magnetic field increases, the diameter of the circles in Figure 6.7 increase. Interesting things happen when a level crosses the Fermi surface.

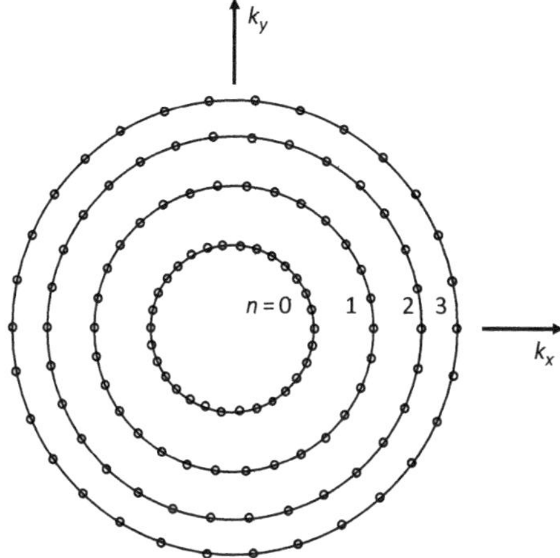

Figure 6.7 Illustration of allowed states in 2D for free electrons in the presence of a uniform magnetic field in the z direction.

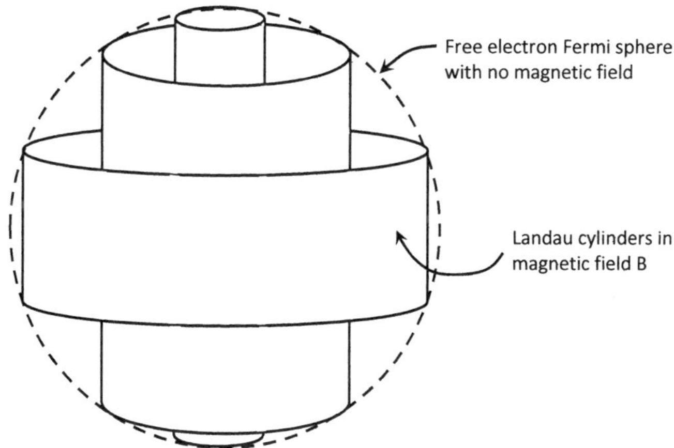

Figure 6.8 The dotted line represents a Fermi sphere for free electrons with $B = 0$. At $T = 0$ all states below the surface of the sphere are occupied; all those above are empty. In the presence of a uniform magnetic field, the possible states are located on the surfaces of concentric cylinders. These cylinders extend infinitely in the $+$ and $-z$ directions. The portions shown are below the Fermi level and occupied. The unoccupied parts of the cylinders, above the Fermi level, are not shown.

A three-dimensional picture is necessary to understand what happens. In analogy to Figure 6.5, in 3D we imagine space to be filled with little cubes with a k state at the corner of each cube. For no magnetic field, the free-electron Fermi surface is the surface of the Fermi sphere. At $T = 0$ all the k states below the Fermi surface are occupied by electrons, all those above are empty. The Fermi surface is indicated by the dotted line in Figure 6.8. With a uniform magnetic field, B, present, the possible states are located on the surface of concentric cylinders as indicated in the figure. These cylinders extend infinitely in the $+$ and $-z$ directions. The portions shown are below the Fermi level and thus occupied. The unoccupied parts, above the Fermi level, are not shown.

Two points are relevant to the behavior as a function of magnetic field: The diameter of the cylinders expands as the field increases; and the number of possible states for a Landau level for each k_z value is proportional to B and independent of n [203]. As the magnetic field increases, Landau levels at higher energies start to empty to lower levels, because the ends of cylinders are moving above the Fermi level, and the degeneracy of these lower levels (number of possible states) is increasing. Examination of Figure 6.8 shows that as a Landau level approaches the Fermi level in a direction perpendicular to the field, an especially large number of states will pass above the Fermi level with a small change in magnetic field. The results of these downward transitions is a lowering of the Fermi level and a lowering of the total energy. However, as the cylinders below the Fermi level continue to expand, the energy is raised and the process repeats. The overall result is an oscillatory behavior of the total energy as a function of magnetic field. The oscillation is periodic in $1/B$ with a period [40, 39]

$$\Delta\left(\frac{1}{B}\right) = \frac{2\pi e}{\hbar A_e}, \tag{6.107}$$

where A_e is any extremal cross-sectional area of the Fermi surface in a direction normal to B. It is not unusual to observe two different extremal cross-sections in the same experiment, which results in beat frequencies. For free electrons, $A_e = \pi k_F^2$. As might be expected, the oscillations in the energy result in oscillations in other physical quantities including magnetization (de Haas van Alphen effect [40, 39]), ultrasonic velocity [3, 198], and ultrasonic attenuation [198, 204, 205]. The oscillations in velocity seem to be observed more frequently than those in the attenuation. A discussion of the coupling between the ultrasonic wave and the electrons, necessary for the oscillations, may be found elsewhere [3, 198].

Acoustic Cyclotron Resonance

Another magnetoelastic effect is that of acoustic cyclotron resonance [1, 2, 197]. This effect occurs when

$$\omega = n\omega_c, \tag{6.108}$$

where ω is the angular frequency of the ultrasonic wave, ω_c is the cyclotron frequency introduced earlier, and n is an integer. The cyclotron frequency is given by $\omega_c = \frac{eB}{m^*}$. The quantity m^* is called the cyclotron effective mass[13] and depends on the shape of the Fermi surface [40]. The attenuation associated with the acoustic cyclotron resonance is periodic in inverse magnetic field with the period,

$$\Delta\left(\frac{1}{B}\right) = \frac{e}{\omega m^*}. \tag{6.109}$$

Thus, such experiments return m^* which is associated with details of the Fermi surface. In order to observe cyclotron resonance one should have $\omega_c \tau > 1$ so that an electron completes at least an orbit before scattering. However in the present case, by Equation 6.108, $\omega \simeq \omega_c$, so the requirement becomes $\omega\tau > 1$. As noted earlier, $\omega\tau = v_s q l_e / v_F = (v_s/v_F) q l_e << q l_e$. Thus the requirement becomes $q l_e >> 1$. This means that acoustic cyclotron resonance imposes difficult conditions, requiring exceptionally pure specimens and low temperatures.

6.9.6 Dislocations

Dislocations are line defects in solids that dramatically affect mechanical properties, decreasing the theoretical strength of the ideal single-crystal by several orders of magnitude. Dislocations occur naturally in the growth process; it is difficult to grow crystals with no dislocations. Dislocations are also produced by mechanical deformation. There are two basic types of dislocations, edge and screw. Illustrations are found in most solid state physics textbooks [39, 40, 203, 41], as well as an explanation as to how they greatly reduce the mechanical strength.

The dislocation is called a line defect because for both the edge and screw dislocation the interatomic bonds are strongly affected mainly in the immediate vicinity of the dislocation line. A single dislocation with no impurities moves with a relatively small applied stress, which is why dislocations reduce the strength of a metal. Metals can be strengthened by making it more difficult to move the dislocation. This can be done by introducing impurities or other defects to "pin" the dislocation. Introducing many more dislocations by deformation results in dislocations crossing each other and very effectively pinning each other. This is the origin of work hardening.

It was recognized very early that dislocations contribute to mechanical dissipation and a reduction of the elastic moduli, and attempts were made to explain these effects [207, 208, 209]. A major advancement was achieved by Granato and Lücke

[13] There are different effective masses in solid state physics depending on the physical effect being discussed, *e.g.*, the specific heat effective mass is usually different than the cyclotron effective mass.

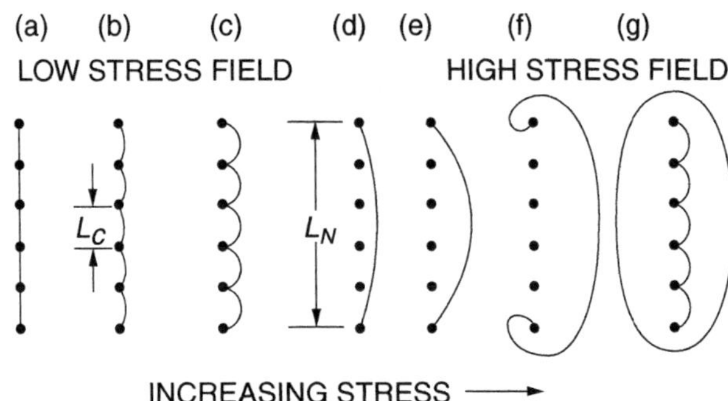

Figure 6.9 An illustration of the vibrating-string model of dislocation damping, showing successive steps in the response of the dislocation to increasing stress. From Granato and Lücke [206]. (a) The undisturbed dislocation. The dots are impurity pinning points, not individual atoms. (b and c) The dislocation line responds to an applied stress, bowing out in response. L_C is the distance between impurity pinning points. The vibrating string model is solved in this region. (d) The string breaks away from the impurity pinning points, but is still constrained by L_N, the network length determined by other dislocations. (e–g) If the stress continues to increase, new dislocations will be formed by the Frank-Reed mechanism.

[206, 210]. Their model treated the dislocation line as a string vibrating under the driving stress of the ultrasonic wave. Figure 6.9 illustrates the basic ideas of their model.

Granato and Lücke [206] solved the damped, vibrating-string model for the dislocation where the strain is due to both the driving ultrasonic wave and the dislocation. For relatively low loss, the solution has a resonance character, but is more complicated than that for a simple, damped harmonic oscillator, being composed of a sum of a series of resonant responses. Figure 6.10 gives the results of their calculation for the first term in the series, the dominant term. The results are given in terms of the decrement, $\Delta = \frac{2\pi v}{\omega}\alpha$, instead of the attenuation α [1]. Further, the results are normalized by terms describing the dislocation including L, the length of the vibrating "string," and ω_o, the "resonant frequency" for the "string." The parameter D is a measure of the loss, the larger D the smaller the loss. In fact, D is just the usual Q. Figure 6.10 gives the results for a very wide of frequencies. For relatively high values of D the system is underdamped and a resonant response is found. For low values of D the system is overdamped and the response is more typical of relaxational attenuation. A more complete treatment of the theory includes a distribution of L_C.

The model has been successful in describing a wide range of phenomena [210]. One particularly nice example involved the irradiation of high-purity copper [211]. The radiation was expected to introduce more pinning defects, thereby reducing the

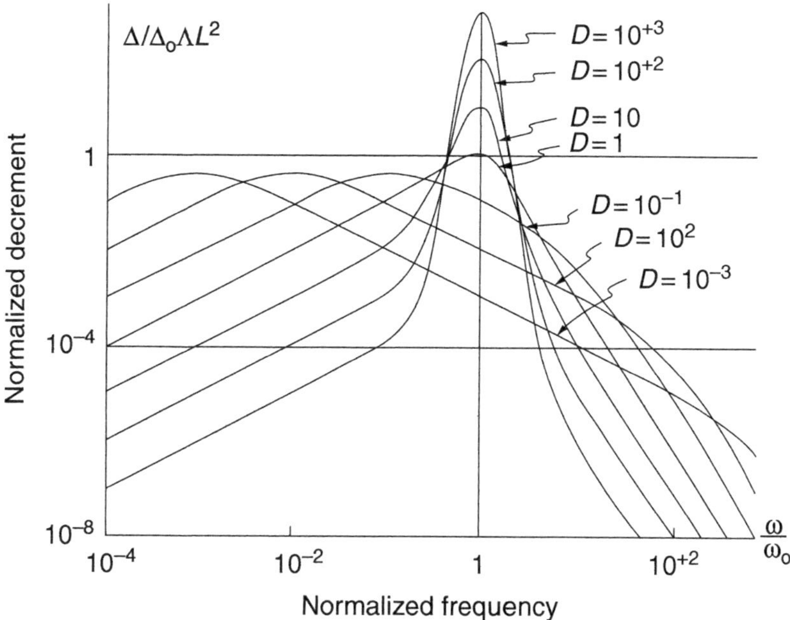

Figure 6.10 Normalized decrement results for various values of D, which is inversely proportional to the damping coefficient. Δ_o is a constant of order unity, Λ is the dislocation density, and L is the loop length. Note the logarithmic scales. From Granato and Lücke [206].

average distance L_C between pinning points. The theory predicts the attenuation to be strongly dependent on L_C, (L_C^4 at low frequencies), and also predicts the resonant frequency $\sim 1/L_C$. The irradiation reduced the magnitude of the attenuation and moved the peak to higher frequencies, just as predicted.

Part (d) of Figure 6.9 shows that at high stress amplitudes of ultrasonic waves the dislocation will breakaway from the pinning points. Such amplitude-dependent effects have indeed observed.

6.9.7 Point Defect Motion

Reference [90] has several chapters devoted to point defects and should be consulted for a more complete coverage than that provided here. "Point" defect refers to an imperfection, composed of only a few atoms, usually one or two, in an otherwise perfect crystal.

Figure 6.11 illustrates a configuration relevant to the Snoek effect [166]. Although important details differ, the Snoek effect exhibits general features common to anelastic relaxation associated with other point defects. A well-known example of the Snoek effect involves a dilute solution of carbon in α iron [212, 90],

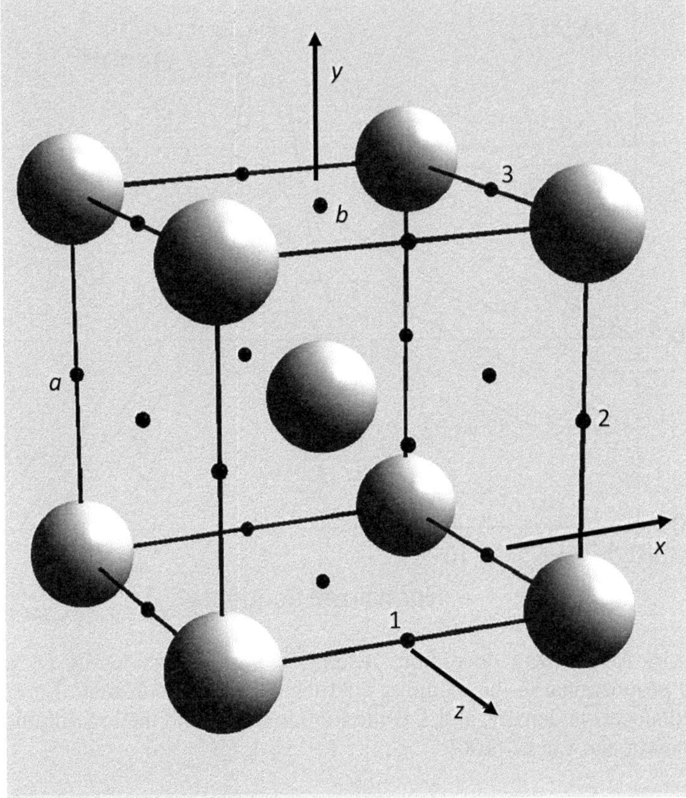

Figure 6.11 One unit cell of a bcc lattice is shown as well as the "octahedral" interstitial sites. The large spheres are the atoms, located at the corners and body center of the cube. The "octahedral" interstitial sites are indicated by the small black dots. There is such a site midway between the atoms located on the corners, and also such a site in the midpoint of each face. Interstitial sites 1, 2, and 3 are different in that the two atoms bracketing each of them are oriented along the x, y, and z directions respectively. Sites a and b are identically situated, because there is another atom above b in the next cube of the lattice (not shown). The sites are called octahedral because the six atoms surrounding each site form an octahedron. The octahedrons extend into adjacent cells. For example, five of the atoms surrounding site b are shown, but the sixth atom lies at the body center of the next cell above.

in which case a small percentage of the interstitial sites in Figure 6.11 are occupied by C and the large atoms are Fe. Obviously, the effect applies to other materials as well. It is easy to see how the relaxational attenuation occurs. Suppose a pure longitudinal wave propagates along the x direction in Figure 6.11. The planes of atoms perpendicular to the x axis will be alternately pushed closer together and pulled further apart. Suppose on compression the atoms on sites type 1 are raised in energy. (The end result is the same if the energy is lowered.) The atoms will readjust their populations to have a larger fraction on sites type 2 and 3. On the expansion cycle, the atoms will tend to move back to sites type 1. This readjustment will occur with a characteristic time τ, leading to relaxational attenuation. From Section 6.6 we have,

$$\frac{1}{Q} = \Delta_R \frac{\omega\tau}{1+\omega^2\tau^2}, \tag{6.110}$$

and

$$\frac{\Delta c}{c} = \frac{c - c_R}{c} = \Delta_R \frac{(\omega\tau)^2}{1+(\omega\tau)^2}. \tag{6.111}$$

Equation 6.111 gives the fractional change in the elastic constant relative to c_R. (This equation sometimes appears in a different form depending on how the fractional change is defined and how Δ_R is defined.) Recalling the definition of Δ_R, Equation 6.77, it is instructive to examine some limits. For $\omega\tau \to \infty$, $c \to c_U$ and for $\omega\tau \to 0$, $c \to c_R$, exactly as expected.

Equations 6.110 and 6.111 should apply rather generally for point defect relaxational attenuation, but τ and Δ_R will depend on the specific defect. These two physical parameters will be examined in turn.

The relaxation time τ is related to τ_{12}, the characteristic time to make a transition between site 1 and site 2, but the two are not equal.[14] An examination of Figure 6.11 shows why. If an interstitial atom occupies a type 1 site, it has several type 2 or type 3 nearest-neighbor interstitial sites to which it could jump, thereby increasing the transition rate. A detailed analysis [90] shows that for the Snoek relaxation $\tau = \tau_{12}/3$. In most cases the factors relating τ to single atom jump rates range form 2–4. In many cases the dependence of τ on some physical parameter, usually temperature, is of more interest than the its exact value.

The parameter Δ_R is usually described in terms of elastic dipoles [90]. Rather than the elastic dipole approach, a simple model, suggested by Ref. [162] will be discussed next. The method will be based on Equation 5.166, which, for convenience, is repeated here

$$\delta c_{mn} = n\left[\frac{1}{Z}\left(\sum_{i=1}^{N}\frac{\partial^2 E_i}{\partial e_m \partial e_n}\exp(-E_i/k_BT) - \frac{1}{k_BT}\sum_{i=1}^{N}\frac{\partial E_i}{\partial e_m}\frac{\partial E_i}{\partial e_n}\exp(-E_i/k_BT)\right)\right.$$
$$\left. + \frac{1}{k_BTZ^2}\left(\sum_{i=1}^{N}\frac{\partial E_i}{\partial e_m}\exp(-E_i/k_BT)\sum_{i=1}^{N}\frac{\partial E_i}{\partial e_n}\exp(-E_i/k_BT)\right)\right]. \tag{6.112}$$

[14] See Nowick and Berry [90] chapter 8 for a full discussion.

It is assumed that there are n independent interstitials atoms distributed randomly among the sites 1, 2, and 3. Then the partition function Z is that for one of the interstitial atoms. The assumption for the strain dependence of the interstitial energy is that

$$E_i = b_i \epsilon_i, \tag{6.113}$$

where b_i is a parameter characterizing the strength of the interaction and the index i runs through 1 to 6. A reasonable assumption is that $b_1 = b_2 = b_3 = b$ and that $b_4 = b_5 = b_6 = 0$; the energy is shifted by compressional strains, but not shear strains. The partition function for a single interstitial atom is

$$Z = \sum_{i=1}^{3} \exp\left(-\frac{E_i}{k_B T}\right) = \sum_{i=1}^{3} \exp\left(-\frac{b\epsilon_i}{k_B T}\right). \tag{6.114}$$

The first example will be for δc_{11}. Making the calculation using Equation 6.112 and then assuming $b\epsilon / k_B T << 1$,

$$\delta c_{11} = -\frac{2}{9}\frac{nb^2}{k_B T}. \tag{6.115}$$

Next, the elastic constant c_{12} will be considered. The calculation is a bit more tedious than for c_{11}, but the result is

$$\delta c_{12} = \frac{1}{9}\frac{nb^2}{k_B T}. \tag{6.116}$$

Of course, $\delta c_{44} = 0$.

Important conclusions may be drawn from these results. Using $B = (c_{11} + 2c_{12})/3$ for the bulk modulus of cubic materials it is seen that $\delta B = 0$. There is no relaxation effect for the bulk modulus. This is just as expected because application of hydrostatic pressure to a cubic material affects all the interstitial sites equally. Another useful shear modulus is $c' = (c_{11} - c_{12})/2$, giving

$$\delta c' = -\frac{1}{6}\frac{nb^2}{k_B T}, \tag{6.117}$$

which is remarkably similar to the result for $\delta c'$ expressed in terms of the force-dipole tensor [213].

From the discussions near Equations 6.77–6.79 we have the relaxation strengths of c_{11} and c' for the Snoek effect,

$$\Delta_R^{11} = \frac{2nb^2}{9c_{11}k_B T}, \tag{6.118}$$

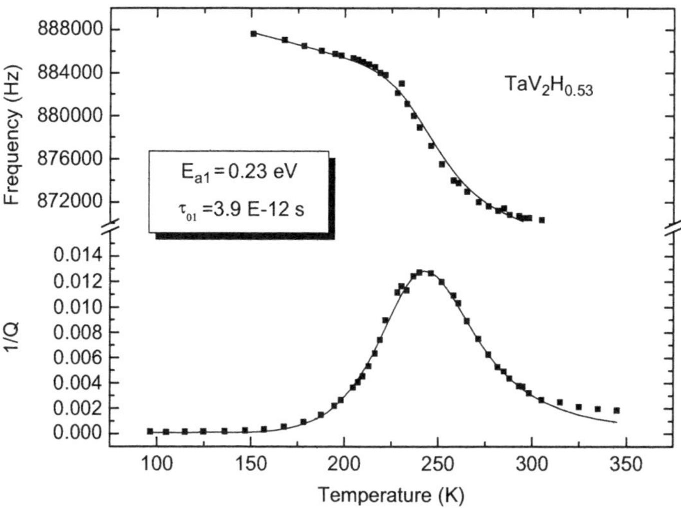

Figure 6.12 Ultrasonic loss and dispersion due to the Snoek effect of H in $TaV_2H_{0.53}$. Exactly the same parameters were used to fit the loss and dispersion (frequency shift), except that a small background, linearly dependent on T, was used for the dispersion [215].

and,

$$\Delta_R' = \frac{nb^2}{6c'k_BT}. \tag{6.119}$$

Interstitial atoms such as C, N, and O dissolved in bbc metals show a Snoek effect. However, the hopping rate of these atoms between interstitial sites is such that the loss peak lies well below ultrasonic frequencies. A light interstitial such as H jumps much faster and is expected to produce an observable loss peak. However, for reasons not entirely clear, H in the elemental bcc metal Ta did not show a loss peak at all [214].

In contrast, well-defined ultrasonic loss peaks were observed, Figure 6.12, in the intermetallic compound C15 TaV_2H_x and attributed to the Snoek effect [215]. This attribution was based on two facts. First, H in C15 TaV_2H_x is known to occupy the tetrahedral g sites in this cubic material, thus by the selection rules of elasticity [90] the Snoek effect is expected. Second, the relaxation strength was linearly dependent on the H concentration, as expected for the Snoek effect, but not for the Zener effect. The linear dependence was observed for six concentrations measured over the range of 0.06 to 0.53.

The network of interstitial sites in TaV_2 is more complicated than in elemental bcc materials. There are 12 g sites per unit formula TaV_2, so the relationship between τ and the single particle hopping time is likely to be somewhat different in the two cases, *and* the expressions for the relaxation strength may differ somewhat, but the differences are not expected to be great. In particular an inverse T dependence is still expected for the relaxation strength.

Figure 6.12 shows an example of the C15 TaV_2H_x results. The measurements were made using RUS (resonant ultrasound spectroscopy). The frequency, f, of a single resonance, which was almost entirely dependent on the polycrystalline shear mode, was measured. In this case

$$\frac{\delta f}{f} = \frac{1}{2}\frac{\Delta c}{c} = \frac{c - c_R}{2c} = \frac{1}{2}\Delta_R \frac{(\omega\tau)^2}{1 + (\omega\tau)^2}. \tag{6.120}$$

The data for the six concentrations were fit with Equations 6.110 and 6.120. Guided by Equations 6.118 and 6.119, the relaxation strength was taken to be

$$\Delta_R = \frac{nD^2}{k_B T}, \tag{6.121}$$

with D an adjustable parameter. The value $D = 0.17$ eV gave a good fit for all six concentrations. Finally, an Arrhenius expression was used for the relaxation time,

$$\tau = \tau_{o1} \exp\left(\frac{E_{a1}}{k_B T}\right), \tag{6.122}$$

with the parameters given in Figure 6.12. It is often the case that a distribution of relaxation times is required to fit the data, but not in this case.

Another well-known example of point defect relaxation is the Zener effect [90, 163] which involves a reorientation of pairs of atoms. A simplified 2D example was discussed in Section 5.8 and illustrated in Figure 5.10. As was the case for the Snoek effect, the loss peak due to Zener relaxation usually occurs at frequencies well below ultrasonic. However, for the light interstitial H, the Zener effect is observed at both low frequencies [164] and ultrasonic frequencies [165, 216, 217], depending on the temperature.

6.9.8 Phase Transitions

Phase transitions were treated in Section 5.7 where static effects were considered, *i.e.*, the results are valid for $\omega \to 0$. However, there are two important dynamic effects to be considered [159, 218, 3]. One is relaxational in nature and the other is related to fluctuations. The ultrasonic attenuation, in particular, is due to dynamic effects. If there is a bilinear coupling between the order parameter and strain

$$F_c = \beta Q e, \tag{6.123}$$

(β a coupling constant, Q an order parameter,[15] and e a strain) then a time-varying strain will produce a time-varying order parameter. In general, there will be a phase

[15] Not to be confused with the $1/Q$ for internal friction.

lag in the response. The result will be relaxational attenuation and Equations 6.110 and 6.111 apply

$$\frac{1}{Q} = \Delta_R \frac{\omega\tau}{1+\omega^2\tau^2}, \tag{6.124}$$

and

$$\frac{\Delta c}{c} = \frac{c - c_R}{c} = \Delta_R \frac{(\omega\tau)^2}{1+(\omega\tau)^2}. \tag{6.125}$$

The ultrasonic attenuation is,

$$\alpha = \frac{\omega}{2v}\left(\frac{1}{Q}\right). \tag{6.126}$$

This description of the ultrasonic response near a phase transition is called the Landau-Khalatnikov theory [219, 159, 218]. (As mentioned earlier, Equation 6.125 sometimes appears in a different form depending on how the fractional change is defined and how Δ_R is defined.) The parameters Δ_R and τ depend on the details of the transition. In particular τ is expected to diverge at the transition, an example of critical slowing down. This situation has been treated in more detail by Fossum [220].

As shown previously, the fluctuation-dissipation theorem connects fluctuations in the stress to ultrasonic attenuation

$$\alpha = \frac{\omega^2 V}{4k_B T \rho v^3} \int_{-\infty}^{\infty} d\tau \, \langle \delta\sigma(0)\delta\sigma(\tau)\rangle \, e^{i\omega\tau}, \tag{6.127}$$

where $\langle \delta\sigma(0)\delta\sigma(t)\rangle$ is a time correlation of the fluctuation stresses and V is the volume of the system. There is, of course an effect on the elastic constant [159, 175]. The fluctuation effects become especially important near phase transitions where the fluctuations of the order parameter become large, and the order parameter is frequently coupled to the strain. Interesting physical results are obtained when the strain (or stress) coupling to the order parameter is inserted into Equation 6.127. Detailed treatments are given for both the linear, ($\simeq eQ$), and quadratic, ($\simeq eQ^2$), cases by both Rehwald [159] and Fossum [220].

6.9.9 Nuclear Acoustic Resonance

Nuclear magnetic resonance (NMR) [221, 222] is a powerful and widely used spectroscopic technique based on the absorption of energy from an oscillating magnetic

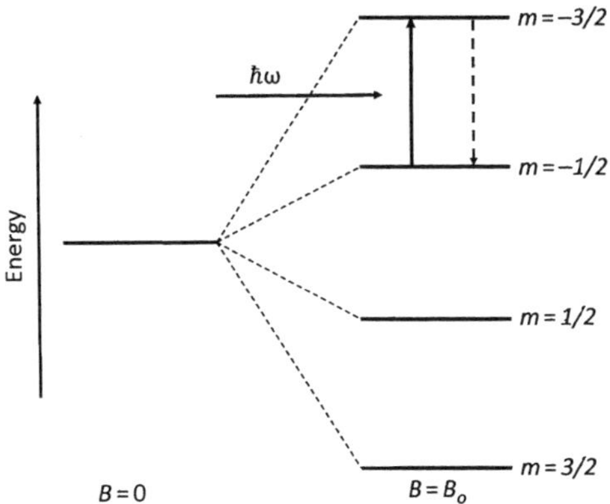

Figure 6.13 Energy levels for a nuclear spin $I = 3/2$ in a magnetic field B. One possible transition is shown. A photon (phonon) is absorbed in an NMR(NAR) experiment producing an upward transition. The nuclear spin may subsequently emit a quantum of energy and make a downward transition.

field. The situation is illustrated in Figure 6.13. A nuclear spin I has quantized energy levels in a magnetic field, B_o. The levels are given by

$$E_m = -m\gamma\hbar B_o, \tag{6.128}$$

where γ is called the gyromagnetic ratio, which is a parameter of the particular nuclei (*e.g.*, it will be different for Cu than for Al), and m is a quantum number which ranges from $+I$ to $-I$ in integer steps. The situation illustrated is for $I = 3/2$. In NMR experiments the nucleus absorbs a quantum of energy from a small magnetic field oscillating at frequency ω. Energy conservation requires $\Delta E = \hbar\omega$ or $\omega = \gamma B_o$. Only $\Delta m = \pm 1$ transitions are allowed. To maintain equilibrium the nucleus may subsequently emit a quantum of energy to the lattice with a characteristic time T_1. The mechanism responsible for the absorption is the interaction of the oscillating magnetic field with the magnetic dipole moment of the nucleus.

The situation differs for nuclear acoustic resonance (NAR) in that $\hbar\omega$ is supplied by externally generated ultrasonic vibrations [87]. Two interactions have been found to produce NAR. The first [86] is due to a dynamic coupling between the time varying electric field gradients produced by the ultrasound, and the nuclear electric quadrupole moment. Both $\Delta m = \pm 1$ and $\Delta m = \pm 2$ transitions are permitted. This mechanism only works for $I > 1/2$ because only such nuclei have electric quadrupole moments.

The second mechanism only works in metals. An acoustic wave propagating in a metal, in the presence of a static magnetic field, produces a transverse current which results in an electromagnetic field oscillating at the acoustic-wave frequency [223]. This effect gives rise to ultrasonic attenuation (Alpher-Rubin effect) [3, 223]. In addition, the coupling of this oscillating field to the nuclear magnetic-dipole moment results in the acoustic excitation of nuclear magnetic resonance [224, 225]. The coupling of the ultrasonic wave to the nuclear spins is similar to that in ordinary NMR (magnetic dipole coupling), thus only $\Delta = \pm 1$ transitions are allowed.

6.9.10 Ultrasonic Paramagnetic Resonance

In an analogous manner to NAR, ultrasonic waves can couple to *electron* spins and produce paramagnetic resonance. This effect has been observed under a number of different experimental conditions [88, 226, 227].

6.9.11 Two-Level Tunneling Systems

Measurement of the specific heat in amorphous insulators yielded a surprise [228]. At low temperatures in crystalline insulators the specific heat is well-described by the Debye model which gives a specific heat varying as T^3. The Debye model holds in the long wavelength limit where the wavelength, λ_{th}, of the thermal phonons is much greater than the interatomic spacing. It was thus expected that the local disorder in amorphous materials would not qualitatively change the thermal phonon contribution so that a T^3 result was expected in amorphous insulators also. Instead, the result for several different amorphous materials behaved as

$$c_v = AT + BT^3. \tag{6.129}$$

The surprise was the extra contribution, AT.

An explanation was soon provided in terms of two-level systems (TLS), which quantum mechanicaly tunnel between two different states [229, 230]. The basic idea is that in disordered materials, some atoms, or small group of atoms, exist in two equilibrium states separated by an energy barrier. The situation is indicated in Figure 6.14. V indicates the potential energy of the tunneling object, an atom or perhaps a small group of atoms. In the absence of tunneling, the energy level difference is ϵ, the asymmetry of the double well. In the cases of interest, the wave functions for the two states may penetrate the barrier, resulting in a tunneling parameter Δ. The Hamiltonian matrix for the system is then,

$$H_o = \frac{1}{2}\begin{bmatrix} \epsilon & \Delta \\ \Delta & -\epsilon \end{bmatrix}. \tag{6.130}$$

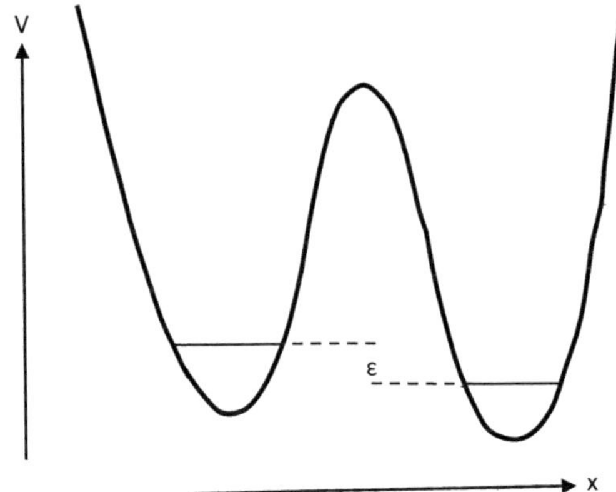

Figure 6.14 An asymetric, double-well potential is illustrated. The lowest energy level in each well is indicated, essentially the lowest vibrational level in each well. In the absence of tunneling, the energy level splitting is ϵ. V indicates the potential energy and x is a configurational coordinate.

Diagonalization of H_o results in two energy levels

$$E_{1,2} = \pm\frac{1}{2}\sqrt{\epsilon^2 + \Delta^2}. \tag{6.131}$$

The energy level splitting is now

$$E = \sqrt{\epsilon^2 + \Delta^2}. \tag{6.132}$$

It is straightforward to calculate the contribution of one such two-level system to the specific heat. To compute the total contribution of the system to the specific heat requires a knowledge of the density of states $\rho(E)$. Under certain assumptions the density of states is taken to be constant. The specific heat then turns out to be linearly dependent on T.

The strain associated with an ultrasonic wave modulates the parameters of the TLS, ϵ and Δ. The result is ultrasonic absorption, both resonance attenuation and relaxational attenuation [231].

Although the TLS model was developed for amorphous insulators, similar effects are found in superionic conductors [232] and metallic glasses [233]. The TLS model has found success in a wide variety of applicatons; however, the microscopic nature of the TLS remains a subject of strong interest [234, 235].

6.9.12 Summary

Some of the loss mechanisms described in this chapter occur in all materials. Most of the others are quite common. Even if one is studying something more exotic, the existing loss mechanisms must be accounted for if reliable results are to be obtained. That necessity was a major motivation for the foregoing survey of common attenuation mechanisms. Less common loss mechanisms, as well as a more detailed treatment of several of the mechanisms treated in this chapter, are to be found elsewhere [1, 2, 3, 90, 169]. It is hoped that the survey presented here will be an encouragement and an aid for deeper study.

Appendix A

Phase Shifts Due to Transducers and Bonds

The concept of acoustic impedance will be useful in the present analysis. In analogy with electrical transmission lines a useful definition of acoustic impedance is [236]

$$Z = -\frac{\sigma}{\dot{u}} \tag{A.1}$$

where σ is the stress and $\dot{u}$ is the associated particle velocity for the acoustic plane wave. The definition is more properly called the specific acoustic impedance or the acoustic impedance per unit area [236, 237]. Consider a longitudinal wave propagating in the x direction, for which the displacement is given by

$$u_x(t) = u_1 \exp\left[i(\omega t - k^* x)\right] + u_2 \exp\left[i(\omega t + k^* x)\right]. \tag{A.2}$$

The first term represents a wave traveling in the positive x direction while the second term represents a wave in the opposite direction. In Equation A.2, $k^* = k - i\alpha$ is the complex propagation constant, and $k = \omega/v$ is the usual wave vector. Then

$$\dot{u}_x = i\omega u_x. \tag{A.3}$$

Using $\sigma_1 = c_{11}\frac{\partial u_x}{\partial x}$,

$$\sigma_1(t) = c_{11} i k^* \left[-u_1 \exp\left[i(\omega t - k^* x)\right] + u_2 \exp\left[i(\omega t + k^* x)\right]\right]. \tag{A.4}$$

Combining the above equations gives

$$Z(x) = \frac{c_{11} k}{\omega}\left[\frac{u_1 \exp(-\gamma x) - u_2 \exp(\gamma x)}{u_1 \exp(-\gamma x) + u_2 \exp(\gamma x)}\right], \tag{A.5}$$

where

$$\gamma = ik + \alpha \tag{A.6}$$

and k^* has been replaced by k, *except* in the exponential. Because $\alpha << k$ this is a good aproximation; but, α is needed in the exponential to express the attenuation of the wave.

Consider the case where there is no reflected wave ($u_2 = 0$); perhaps the right end of the specimen is perfectly absorbing. Then Equation A.5 gives

$$Z(x) = \frac{c_{11}k}{\omega} = \frac{(\rho v^2)\omega/v}{\omega} = \rho v \equiv Z_o. \tag{A.7}$$

Z_o is known as the characteristic impedance of the material. Two more steps lead to the desired equation for transformation of impedances along the direction $\hat{x}$. Suppose that at the point $x = l$ the specimen is terminated in some arbitrary impedance $Z(l)$. Applying Equation A.5,

$$Z(l) = Z_o \left[\frac{u_1 \exp(-\gamma l) - u_2 \exp(\gamma l)}{u_1 \exp(-\gamma l) + u_2 \exp(\gamma l)} \right]. \tag{A.8}$$

Similiarly, calling Z_{in} the impedance at $x = 0$ gives,

$$Z_{in} = Z_o \frac{u_1 - u_2}{u_1 + u_2}. \tag{A.9}$$

Equations A.8 and A.9 may be used to eliminate u_1 and u_2, with the result,

$$Z_{in} = Z_o \frac{Z(l) + Z_o \tanh(\gamma l)}{Z_o + Z(l) \tanh(\gamma l)} \tag{A.10}$$

Equation A.10 is the well known, and very useful, equation for the transformations of impedances on either electrical or acoustic transmission lines [238].

It will also be useful to have expressions for transmission and reflection coefficients for the displacements and stresses. The situation is analyzed with the help of Figure A.1. A displacement wave of amplitude A_1 is incident from the left on a boundary betweeen materials i and j. Part of the wave is transmitted through the interface and part is reflected. The situation for both displacements and stresses is shown. It is assumed that the displacements and stresses are continuous at the boundary. Requiring the displacements to be continuous at the boundary yields

$$A_1 + B_1 = A_2, \tag{A.11}$$

and requiring the stress to be continuous gives

$$-Z_iA_1 + Z_iB_1 = -Z_jA_2, \tag{A.12}$$

where $k_ic_i = \frac{\omega}{v_i}\rho v_i^2 = \omega\rho v_i = \omega Z_i$, has been used. Z_i is the characteristic impedance for medium i.

It is desired to obtain the reflection coefficient $R_{ij} = B_1/A_1$ and the transmission coeffcent $T_{ij} = A_2/A_1$; these are both for the *displacements*. Solving Equations A.11 and A.12 results in

$$R_{ij}^d = \frac{Z_i - Z_j}{Z_i + Z_j} \qquad T_{ij}^d = \frac{2Z_i}{Z_i + Z_j}, \tag{A.13}$$

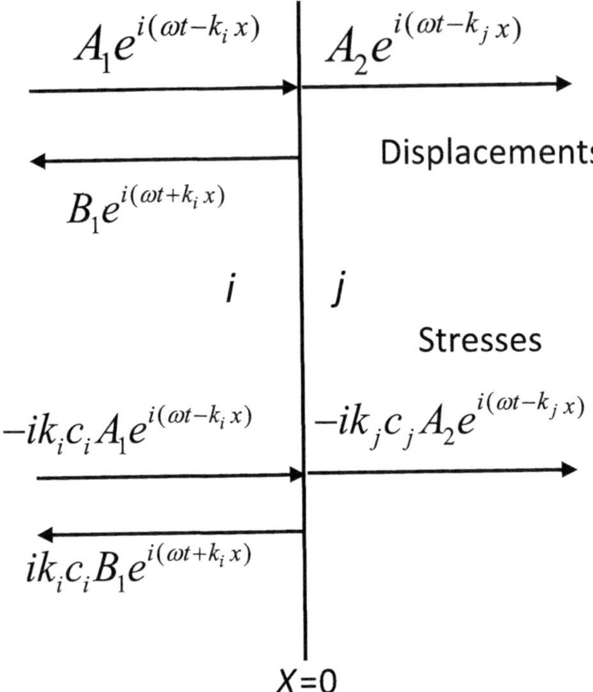

Figure A.1 Reflection and transmission of waves at a boundary between materials i and j. The equations for the displacements are given in the top part and the corresponding equations for the stresses are given below. The c_i and c_j represent the elastic constants in the two materials.

where the superscript d is a reminder that these coefficients are for the displacement. Proceeding similarly for the stress gives

$$R^s_{ij} = \frac{Z_j - Z_i}{Z_i + Z_j} \qquad T^s_{ij} = \frac{2Z_j}{Z_i + Z_j}. \tag{A.14}$$

The formalism just reviewed can be used to calculate the phase shift ϕ, and hence $\Delta t(\phi)$, of Equation 4.7. Referring to Figure A.2, suppose it is desired to measure the time interval between successive echoes, *e.g.*, 1 and 2. Echo 2 results from an extra round-trip through the specimen as compared to echo 1. The phase shift due to the reflection at the transducer, *i.e.*, between A and B, is needed. Equation A.13 will be used, but that requires Z_B and *that* requires Z_T. Equation A.10 is the needed tool, but following McSkimin [77, 78], the loss in the transducer and bond will be neglected. With no loss $\gamma = ik$ and Equation A.10 becomes

$$Z_{in} = Z_o \frac{Z(l) + iZ_o \tan(kl)}{Z_o + iZ(l) \tan(kl)}. \tag{A.15}$$

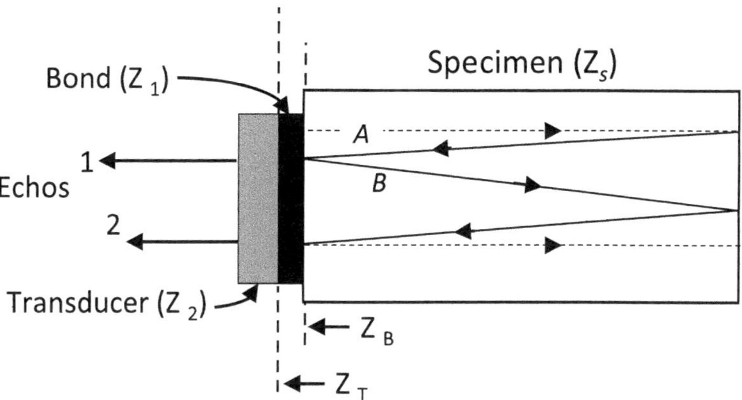

Figure A.2 Illustrations of paths resulting in two successive echoes. The arrows shown within the specimen are only to illustrate the different round trips and reflections, and, of course do not represent the actual paths of the plane waves. The thicknesses of the transducer and bond are exaggerated for clarity. Z_1, Z_2, and Z_S are the characteristics impedances of the bond, transducer, and specimen, respectively. The path resulting in echo 2 undergoes an extra reflection at the transducer-bond interface (as well as an extra reflection at the free end of the specimen). Z_T is the input impedance of the transducer; Z_B is the input impedance of the bond-transducer combination. The phase change between A and B is needed.

A quick application of Equation A.15 produces

$$Z_T = iZ_2 \tan(k_T l_T), \tag{A.16}$$

where k_T and l_T are for the transducer.[1] Another quick application of Equation A.15 produces

$$Z_B = iZ_1 \frac{Z_2 \tan(k_T l_T) + Z_1 \tan(k_B l_B)}{Z_1 - Z_2 \tan(k_T l_T) \tan(k_B l_B)}. \tag{A.17}$$

The subscript B refers to the bond. Let A and B represent the amplitude of the *displacement* waves involved in the indicated reflection at the bond-specimen interface. Then

$$\frac{B}{A} = R^d_{SB} = \frac{Z_S - Z_B}{Z_S + Z_B}. \tag{A.18}$$

The displacement reflection coefficient is put into a more convenient form by writing[2]

$$R^d_{SB} = \exp(i\phi), \tag{A.19}$$

[1] It is worth noting that $l_T = \lambda/2$ and at transducer resonance, $k_T l_T = \pi$ and $Z_T = 0$.

[2] For an arbitrary complex number, there would be a coefficient before the exponential term, but for the special form of Equation A.18, this coefficient is unity.

where

$$\begin{aligned} \phi &= \arctan\left[\frac{\mathrm{Im}(R^d_{SB})}{\mathrm{Re}(R^d_{SB})}\right] && \text{if Re}\left(R^d_{SB}\right) \geq 0 \\ \phi &= \arctan\left[\frac{\mathrm{Im}(R^d_{SB})}{\mathrm{Re}(R^d_{SB})}\right] + \pi && \text{if Re}\left(R^d_{SB}\right) < 0. \end{aligned} \tag{A.20}$$

If at the left end of the specimen $A = A_o \exp(i\omega t)$, then $B = \exp(i\phi)A_o \exp(i\omega t) = A_o \exp(i(\omega t + \phi))$. If at a time t_A, A achieves a certain amplitude, then B reaches *the same amplitude* at a time t_B given by $\omega t_A = \omega t_B + \phi$. The difference between these two times $t_B - t_A = -\phi/\omega$ leading to

$$\Delta t(\phi) = t_B - t_A = -\frac{\phi}{\omega} = -\frac{\phi}{2\pi f}. \tag{A.21}$$

This difference in the times is used in Equation 4.7, except in the latter case provision is made for p round trips through the specimen by the second echo.

(The preceding analysis *could* have been carried out using the *stress* reflection coefficient, Equation A.14, rather than the *displacement* coefficient, Equation A.13. There appears to be a problem, because these two equations have opposite signs. However, the *stress* coefficient produces a sign change upon the reflection of wave B at the right end of the specimen, while the *displacement* coefficient does not. The phase difference between echoes 1 and 2 would be the same in either case.)

In summary, the phase shift ϕ can be calculated if the appropriate parameters are available. These parameters are ρ and v for the specimen,[3] bond material, and transducer. Also needed are the thicknesses of the bond and transducer.

[3] The phase shift ϕ involves a *correction* term of the order of a few percent, so an approximate value of the specimen velocity should be sufficient at this point.

Appendix B

Diffraction

Plane-wave propagation is, of course, an idealization. In reality, a beam from a finite aperture (the transducer) will undergo beam spreading (diffraction). It is obvious that the diffraction means a decrease in the energy impinging on the receiving transducer and hence a contribution to the loss. It is perhaps less obvious that the phase of the received wave advances due to diffraction. In the discussion of diffraction effects it is convenient to define the parameter $S = z\lambda/a^2$ where a is the radius of the transducer, λ is the wavelength of the ultrasound, and z is the distance the ultrasound travels between the transmitting and receiving transducer(s) (often the same transducer as in Figure 4.2). For the one-transducer case $z = p(2L)$ where p is the number of round trips.

Treating the transducer as a piston source, the effects of diffraction have been calculated [236, 239, 240]. Figure B.1 gives results for an isotropic material where the phase advance and the loss are shown as a function of S. (The anisotropic has also been treated [239, 240].) Consider two echoes, m and $m + p$. From Figure B.1 it is seen that the t_D of Equation 4.7 is given by

$$t_D = \frac{\phi(S_{m+p}) - \phi(S_m)}{2\pi f}. \tag{B.1}$$

It seems worth noting that the total phase advance from $S = 0$ to infinity is $\pi/2$, thus the maximum value of t_D is $1/(4f)$. It seems quite feasible to design experiments such that $\Delta S \approx 0.1$ to 0.2 radians, giving a range of $t_D \approx 1/60f$ to $1/30f$. Obviously, if it is possible to choose $a >> \lambda$ diffraction effects are minimized.

In addition to the above considerations, it is advisable to have the lateral dimensions of the specimen considerably greater than the diameter of the transducer; otherwise, sidewall reflections of the diffracted beam may mix back into the main undiffracted beam with resulting complications [1].

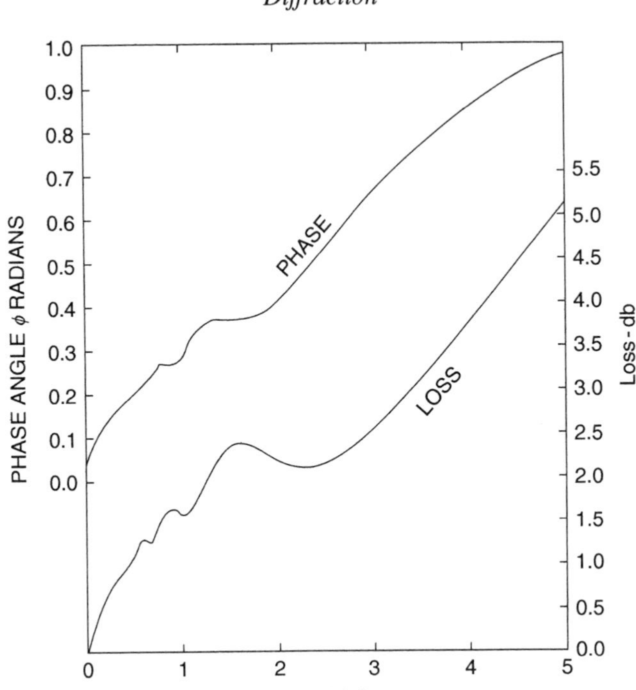

Figure B.1 The calculated phase advance and the ultrasonic loss due to diffraction from a piston source for an isotropic material. (From Papadakis [239]). The parameter S is discussed in the text.

Appendix C

Transducer Effects on Resonant Frequencies

The treatment of the 1D plane-wave oscillator given in Section 4.1.2 was for an isolated oscillator. In practice the attached transducer(s) will affect the measured resonant frequencies. Using the methods introduced in Appendix A, it is straightforward, if a bit tedious, to treat this problem. The situation is illustrated in Figure C.1. Transducers are attached to opposite ends of the specimen. The bonds are assumed to be thin enough to be ignored. The parameters Z_i, l_i, v_i are the characteristic impedances, lengths, and sound velocities for the transducer and specimen. The approach will be to determine Z_{in} in terms of the given parameters; then, set $Z_{in} = 0$ to determine the resonant frequencies of the composite resonator. Multiple applications of Equation A.15 should do the job.[1]

$$Z_1 = iZ_{T1} \tan(k_{T1} l_{T1}), \tag{C.1}$$

$$Z_2 = Z_s \frac{Z_1 + iZ_s \tan(k_s l_s)}{Z_s + iZ_1 \tan(k_s l_s)}, \tag{C.2}$$

and

$$Z_{in} = Z_{T2} \frac{Z_2 + iZ_{T2} \tan(k_{T2} l_{T2})}{Z_{T2} + iZ_2 \tan(k_{T2} l_{T2})}. \tag{C.3}$$

Combining Equations C.1, C.2, and C.3 and setting $Z_{in} = 0$ gives the requirement for the resonance [69],

$$\begin{aligned} Z_s Z_{T1} \tan(k_{T1} l_{T1}) + Z_s Z_{T2} \tan(k_{T2} l_{T2}) + Z_s^2 \tan(k_s l_s) \\ - Z_{T1} Z_{T2} \tan(k_{T1} l_{T1}) \tan(k_{T2} l_{T2}) \tan(k_s l_s) = 0. \end{aligned} \tag{C.4}$$

[1] In calculating this correction term for the measured frequency, the loss in the specimen and transducers is ignored.

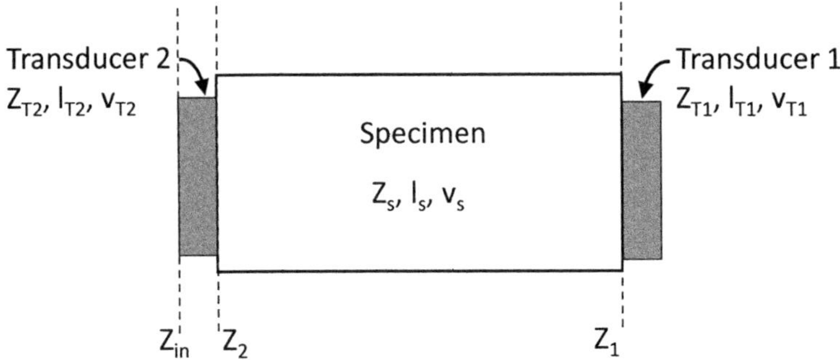

Figure C.1 Transducers are attached to opposite ends of the specimen. The characteristic impedance Z_i, thickness l_i, and sound velocity v_i are labeled for the specimen and transducers. Z_1 is the impedance observed looking into the transducer on the right; Z_2 is the impedance seen looking into the specimen from its left end; and Z_{in} is the input impedance for the composite resonator.

Equation C.4 can be put into a more convenient form for the present case by the following considerations. From Equation 4.11 $f_n^s = nv_s/2l_s$, where f_n^s indicates the nth resonance of the isolated specimen. Then

$$k_s l_s = \left(\frac{2\pi f}{v_s}\right)\left(\frac{nv_s}{2f_n^s}\right) = \frac{n\pi f}{f_n^s} = n\pi\left(\frac{f - f_n^s}{f_n^s}\right) + n\pi. \tag{C.5}$$

Similar considerations apply to the transducer, $l_T = \lambda_T/2 \rightarrow f^T = v_T/2l_T$, leading to

$$k_T l_T = \pi\left(\frac{f - f^T}{f^T}\right) + \pi, \tag{C.6}$$

where this relation holds for either transducer. Two different situations are now considered.

Identical Transducers on Each End

The two transducers are identical. Setting $f = f_n^C$, where f_n^C designates the nth resonance of the composite transducer-specimen assembly, assuming $\left(f_n^C - f_n^s\right)/f_n^s << 1$ and $\left(f_n^C - f_T\right)/f^T << 1$, using the identity that $\tan(\theta + n\pi) = \tan(\theta)$, the tangents in Equation C.4 will be replaced by their arguments. Setting $T1 = T2 = T$

$$2Z_s Z_T \pi\left(\frac{f_n^C - f^T}{f^T}\right) + Z_s^2 n\pi\left(\frac{f_n^C - f_n^s}{f_n^s}\right) - Z_T^2 n\pi^3\left(\frac{f_n^C - f^T}{f^T}\right)^2\left(\frac{f_n^C - f_n^s}{f_n^s}\right) = 0 \tag{C.7}$$

Defining $\eta = (Z_T/Z_s)\left(f_n^s/\left(nf^T\right)\right)$, the above equation can be rewritten

$$\left(f_n^C - f_n^s\right) + 2\eta\left(f_n^C - f^T\right) - \eta^2\left(n\pi/f_n^s\right)^2\left(f_n^C - f^T\right)^2\left(f_n^C - f_n^s\right) = 0. \tag{C.8}$$

It is important to note

$$\eta = (Z_T/Z_s)\big(f_n^s/\big(nf^T\big)\big) = \frac{\rho_T l_T}{\rho_s l_s}. \tag{C.9}$$

In practical applications $\eta \approx 1\%$ or much less for thin film transducers.

The third term in Equation C.8 is much smaller than the second term, and will be neglected, with the result

$$f_n^s = f_n^C + 2\eta\left(f_n^C - f^T\right), \tag{C.10}$$

and

$$f_{n+1}^s - f_n^s = \Delta f^s = \left(f_{n+1}^C - f_n^C\right)(1 + 2\eta). \tag{C.11}$$

Equations C.10 and C.11 provide the needed corrections for the transducer effects on the *measured* resonant frequency, f_n^C, giving Δf^s the frequency of the isolated specimen which is just $2l/v_s$. Higher-order corrections have been discussed in Refs. [83] and [69].

Single Transducer

The case of a single transducer is easily treated by setting $l_{T1} = 0$ in Equation C.4. The results are simple

$$f_n^s = f_n^C + \eta\big(f_n^C - f^T\big), \tag{C.12}$$

and

$$f_{n+1}^s - f_n^s = \Delta f^s = \left(f_{n+1}^C - f_n^C\right)(1 + \eta). \tag{C.13}$$

Appendix D

Damped, Driven Oscillator and Complex Force Constant

A damped, driven oscillator was used in Section 4.2.3 as a model for a single RUS resonance, thereby allowing for a theoretical fit to the experimental resonance curves for the purpose of a sensitive determination of Q and f_n. The damping was taken to be a velocity-dependent force $-b\frac{dx}{dt}$. While such a force is appropriate for certain motions in fluids, it is not so clear that this is a good model for solids. The use of a complex force constant allows for more flexibility. In this case the equation of motion becomes

$$Fe^{-i(2\pi ft+\theta)} - (k_1 - ik_2)x = m\frac{d^2x}{dt^2}, \tag{D.1}$$

where k_1 and k_2 are the real and imaginary parts of the force constant. Setting

$$x(t,f) = L(f)e^{-i2\pi ft}, \tag{D.2}$$

then

$$L(f) = A\frac{e^{-i\theta}}{(f_n^2 - f^2) - i(k_2/((2\pi)^2m)}. \tag{D.3}$$

Different physical situations correspond to different forms of k_2. Several relevant terms are defined in Section 4.3.2.

For the first case, consider $k_2/m = (2\pi f_n)^2/Q$, independent of frequency. Then

$$L1(f) = A\frac{e^{-i\theta}}{(f_n^2 - f^2) - i(f_n^2/Q)}, \tag{D.4}$$

which is almost, but not quite, the same as Equation 4.44. (f_n is the *nth* resonant frequency as described in Section 4.2.3.)

For the second case, consider $k_2/m = (2\pi f_n)(2\pi f)/Q$, linearly dependent on f. Then

$$L2(f) = A\frac{e^{-i\theta}}{(f_n^2 - f^2) - i(f_n f/Q)}, \tag{D.5}$$

which is exactly the same as Equation 4.44. Near f_n Equations D.4 and D.5 give almost identical results and it doesn't matter which one is used to fit the resonance line. Far off resonance the two results differ substantially, which is important for thermal noise in certain mechanical oscillators used in gravitational wave detectors [241]. The first case corresponds to the internal friction $1/Q$ being independent of frequency. The second case corresponds to the internal friction being linearly dependent on frequency.

Appendix E

Comparison of the Quasistatic and Experimental Temperature Dependence for Specific Cases

In this Appendix the quasistatic approximation will be tested by applying Equations 5.72 to two specific cases: silver and diamond. For these examples the $\bar{c}_{ijkl}$ values, *i.e.*, initial values, will be taken to be the experimental 0 K results. The expansion due to the zero point motion is included in these parameters.

E.1 Silver

Table E.1 gives the third-order elastic constants for silver and also diamond, which will be discussed in the follwing example. For Ag, the experimental second-order elastic constants are found in Refs. [242] and [27]. Also needed to carry out the calculations is ξ, the fractional change in length due to thermal expansion [243]. (The inset figure on p. 299 of Ref. [243] has an error. The 10^{-6} on the vertical axis should be instead 10^{-5}). Figure E.1 is based on the data of Ref. [243].

The third-order elastic constants, the 0 K second-order elastic constants, and the fractional change in length due to thermal expansion will now be used in Equations 5.72 to calculate the temperature-dependent, second-order elastic constants of Ag. The calculated results for c_{11} are shown in Figure E.2 along with the experimental results.

Table E.1 *Third order elastics constants (GPa).*

	Silver [244]	Diamond [245]
c_{111}	−843	−7603
c_{112}	−529	−1909
c_{123}	189	835
c_{144}	56	1438
c_{166}	−637	−3938
c_{456}	83	−2316

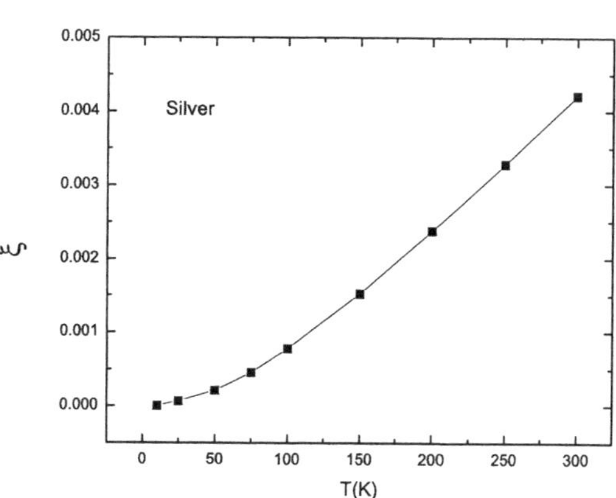

Figure E.1 Fractional change in length for silver over the temperature range of 10–300 K. Based on the data of Ref. [243].

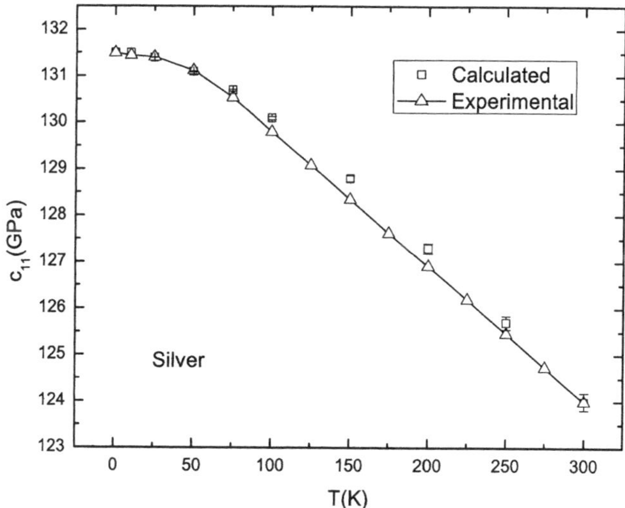

Figure E.2 c_{11} for silver. The calculated values are from Equation 5.72. The experimental values are from Refs. [242, 27].

Figure E.3 shows similar results for c_{12} and c_{44}. Figures E.2 and E.3 show exceptionally close agreement between the experimental values and those calculated based on thermal expansion and third-order elastic for all three independent elastic constants for silver. Error bars resulting from errors in the measurements of the third-order elastic constants were calculated for Figures E.2 and E.3. These are barely visible in the figures. The calculated and experimental results are in good agreement, and provide strong support for the quasistatic approximation, at least for silver.

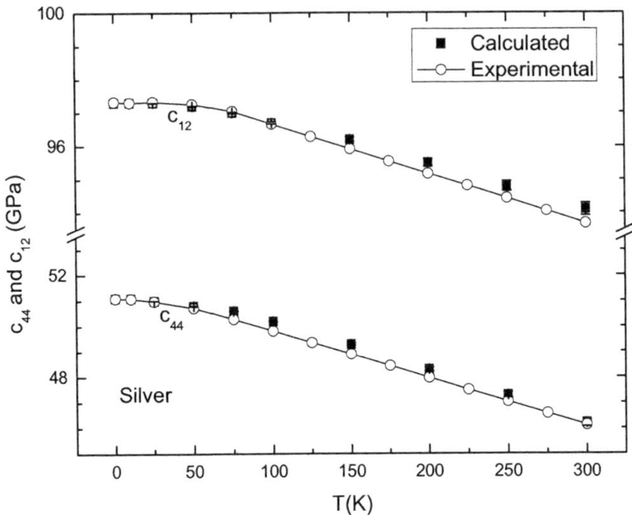

Figure E.3 c_{12} and c_{44} for silver. The calculated values are from Equations 5.72. The experimental values are from Refs. [242] and [27].

E.2 Diamond

As a second example, a very different material is considered: single-crystal diamond. The third-order elastic constants [245], the thermal expansion [246], and the temperature-dependent second-order elastic constants have all been measured [151]. Figure E.4 shows the fractional change in length for diamond over the temperature range of 0–300 K. The data for this graph were obtained by a numerical integration of figure 5 of Ref. [246]. Comparison of Figures E.1 and E.4 show that diamond has a much smaller change in length than silver, and the temperature dependence is quite different.

Figure E.5 gives the calculated and experimental values for c_{11} of diamond. Figure E.6 compares calculated and experimental results for c_{12} and c_{44} of diamond. Error bars resulting from errors in the third-order elastic constants were computed, and are shown in the figures. For both the diamond c_{11} and c_{44} the computed and experimental results are in reasonably good agreement, although there appears to be a small discrepancy for the case of c_{11}. For c_{12} the difference between computed and experimental values is larger than for the other cases, but the error bars are considerably larger in this case, primarily due to the error in c_{123}. For the six elastic constants considered, three for silver and three for diamond, there is no significant difference between computed and measured values.

The third-order elastic constants used in the calculations were measured at room temperature. At least in a few cases [247, 248], some of the third-order elastic

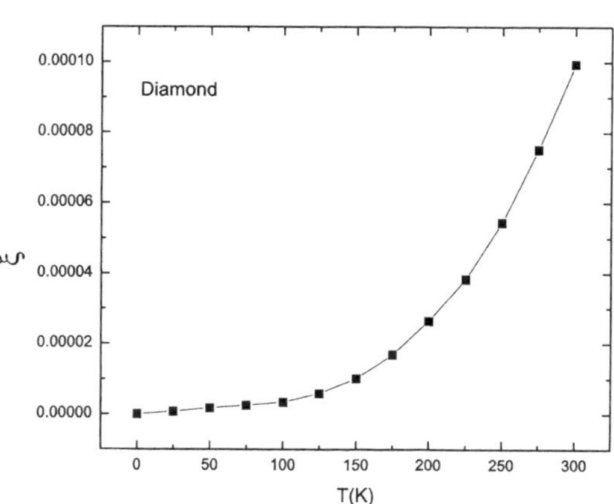

Figure E.4 Fractional change in length for diamond over the temperature range of 10–300 K. The data for this graph were obtained by a numerical integration of figure 5 of Ref. [246].

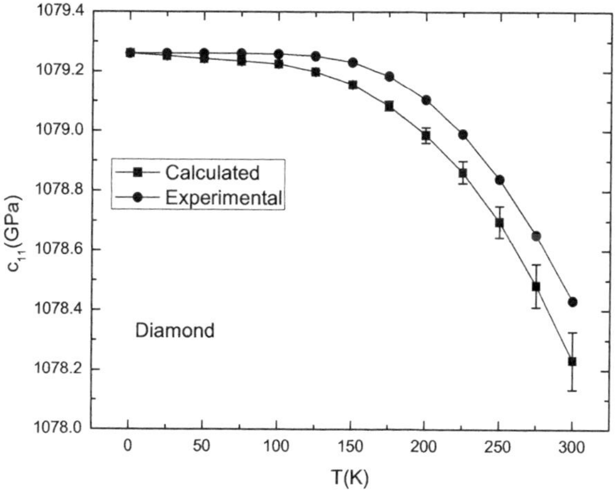

Figure E.5 c_{11} for diamond. The calculated values are from Equation 5.72. The experimental values are from Ref. [151]. The error bars are calculated from the data given in Ref. [245].

constants depend on temperature. Some improvement in the agreement between computed and experimental values could probably be achieved by taking such temperature dependence into account in cases where the information is available. The variation in the third-order elastic constants appears to occur at low temperatures. The smallness of the thermal expansion coefficients at these temperatures mitigates the effect of this variation on the calculated second-order elastic constants.

Overall, these results provide strong support for the idea that the temperature dependence of the elastic constants for many materials is largely explained by the

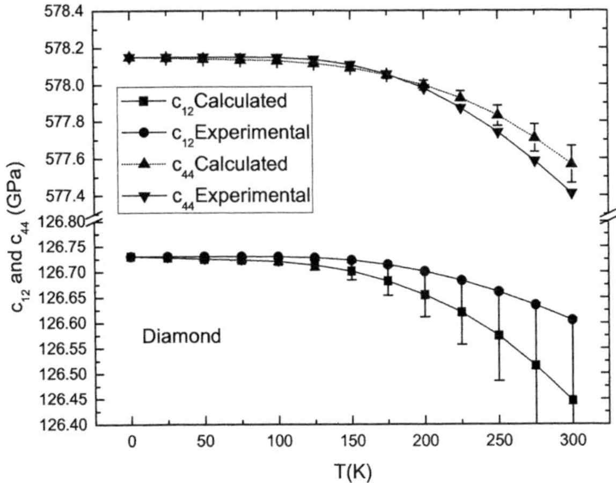

Figure E.6 c_{12} and c_{44} for diamond. The calculated values are from Equation 5.72. The experimental values are from Ref. [151]. The error bars are calculated from the data given in Ref. [245].

quasistatic approximation. The situation will likely be more complicated in many cases. Near a phase transition, for example, it seems reasonable to expect the third-order elastic constants may depend strongly on temperature, which would need to be taken into account.

References

[1] R. Truell, C. Elbaum, and B.B. Chick. *Ultrasonic methods in solid state physics*. Academic Press, 1969.

[2] R.T. Beyer and S.V. Letcher. *Physical ultrasonics*. Academic Press, 1969.

[3] B. Lüthi. *Physical acoustics in the solid state*. Springer, 2005.

[4] J.F. Nye. *Physical properties of crystals*. Oxford, 1979.

[5] L.D. Landau and E.M. Lifshitz. *Theory of elasticity*. 3rd ed. Butterworth Heinemann, 1986.

[6] F.I. Fedorov. *Theory of elastic waves in crystals*. Oxford, 1986.

[7] D.C. Wallace. *Thermodynamics of crystals*. Wiley, 1972.

[8] R.L. Melcher. "Physical acoustics." In: ed. by W.P. Mason and R.N. Thurston Vol. XII. Academic Press, 1976. Chap. 1.

[9] G. Arfken. *Mathematical methods for physicists*. 3rd ed. Academic Press, 1985.

[10] K.D. Swartz and A.V. Granato. "Experimental test of the Laval Raman Viswanathan theory of elasticity." In: *J. Acoust. Soc. Amer* 38 (1965), p. 824.

[11] B.A. Auld. *Acoustic fields and waves in solids*. Second. Vol. 1. Krieger, 1990.

[12] W.A. Wooster. *A textbook on crystal physics*. Cambridge University Press, 1938.

[13] W. Voigt. *Lehrbuch der kristallphysik*. Johnson Reprint Corporaton, 1966.

[14] M. Levy. "Handbook of Elastic Properties of Solids, Liquids, and Gases." In: ed. by M. Levy, H.E. Bass, and R.R. Stern. Vol. II. Academic Press, 2001. Part 1, Chapter 1.

[15] D.B. Litvin. "The icosahedral point groups." In: *Acta Cryst.* 47 (1991), p. 70.

[16] P.S. Spoor. "Elastic properties of novel materials using PVDF film and resonance ultrasound spectroscopy." PhD thesis. Pennsylvania State University, 1997.

[17] M.A. Chernikov, H.R. Ott, A. Bianchi, A. Migliori, and T.W. Darling. "Elastic moduli of a single quasicrystal of decagonal Al-Ni-Co: Evidence for transverse elastic isotropy." In: *Phys. Rev. Lett* 80 (1998), p. 321.

[18] D. Levine, T.C. Lubensky, S. Ostlund, S. Ramaswamy, P.J. Steinhardt, and J. Toner. "Elasticity and dislocations in pentagonal and icosahedral quasicrystals." In: *Phys. Rev. Lett.* 54 (1985), p. 1520.

[19] Y. Ishii. "Phason softening and structural transitions in icosahedral quasicrystals." In: *Phys. Rev. B* 45 (1992), p. 5228.

[20] M. Oxborrow and C.L. Henley. "Random square-triangle tilings." In: *J. Non-Cryst. Solids* 153 (1993), p. 210.

[21] M. Oxborrow and C.L. Henley. "Random square-triangle tilings: A model for twelvefold-symmetric quasicrystals." In: *Phys. Rev. B* 48 (1993), p. 6966.

[22] K. Foster, S.L. Fairburn, R.G. Leisure, S. Kim, D. Balzar, G. Alers, and H. Ledbetter. "Acoustic study of texture in polycrystalline brass." In: *Acoust. Soc. Amer* 105 (1999), pp. 2663–2668.

[23] R. Lakes. "Foam structures with a negative poissons ratio." In: *Science* 235 (1987), p. 1038.

[24] R.F.S. Hearmon. *An introduction to applied anisotropic elasticity*. Oxford University Press, 1961.

[25] S.P. Timoshenko and J.N. Goodier. *Theory of elasticity*. McGraw-Hill, 1970.

[26] S.G. Lekhnitskii. *Theory of elasticity of an anisotropic elastic body*. Holden-Day, 1963.

[27] G. Simmons and H. Wang. *Single crystal elastic constants and calculated aggregate properties: a handbook*. MIT Press, 1971.

[28] H. Ledbetter. "Handbook of elastic properties of solids, liquids, and gases." In: ed. by M. Levy and L. Furr. Vol. III. Academic Press. Chap. 11, p. 313.

[29] W.C. Cady. *Piezoelectricity*. Vol. One. Dover, 1964.

[30] J.R. Neighbours and G.E. Schacher. "Determination of elastic constants from sound-velocity measurements in crystals of general symmetry." In: *J. Appl. Phys.* 38 (1967), p. 5366.

[31] H. Ledbetter, R.G. Leisure, A. Migliori, J. Betts, and H. Ogi. "Low-temperature elastic and piezoelectric constants of paratellurite (α-TeO_2)." In: *J. Appl. Phys* 96 (2004), p. 6201.

[32] E.B. Christoffel. "Uber die Fortpflanzung von Stossen durch elastische feste Korper." In: *Ann. Mat. Pura Appl.* 8 (1877), p. 193.

[33] D. Royer and E. Dieulesaint. *Elastic waves in solids I*. Springer, 1996.

[34] E.S. Fisher and C.J. Renken. "Single-crystal elastic moduli and the hcp bcc transformation in Ti, Zr, and Hf." In: *Phys. Rev.* 135 (1964), A482.

[35] M.H. Manghnani. "Elastic constants of single-crystal rutile under pressures to 7.5 kilobars." In: *J. Geophy. Res.* 74 (1969), p. 4317.

[36] W.J. Alton and A.J. Barlow. "Acoustic-wave propagation in tetragonal crystals and measurement of elastic constants of calcium molybate." In: *J. Appl. Physics.* 38 (1967), p. 3817.

[37] H.J. McSkimin. "Temperature dependence of the adiabatic elastic moduli of single-crystal alpha uranium." In: *J. Appl. Phys.* 31 (1960), p. 1627.

[38] J.D. Jackson. *Classical electrodynamics*. Wiley, 1999.

[39] C. Kittel. *Introduction to solid state physics*. 8th edn. Wiley, 2005.

[40] N.W. Ashcroft and N.D. Mermin. *Solid state physics*. Saunders, 1976.

[41] M.P. Marder. *Condensed matter physics*. Wiley, 2010.

[42] G. Burns. *Solid state physics*. Academic Press, 1985.

[43] M. Born and K. Huang. *Dynamical theory of crystal lattices*. Clarendon Press, 1954.

[44] H. Böttger. *Principles of the theory of lattice dynamics*. Physik-Verlag, 1983.

[45] M.T. Dove. *Introduction to lattice dynamics*. Cambridge University Press, 1993.

[46] B.T.M. Willis and A.W. Pryor. *Thermal vibratons in crystallography*. Cambridge University Press, 1975.

[47] W. Cochran. "Lattice vibrations." In: *Rep. Prog. Phys.* 26 (1963), p. 1.

[48] F. Herman. "Lattice vibrational spectrum of germanium." In: *J. Phys. Chem. Solids.* 8 (1959), p. 405.

[49] G.P. Srivastava. *The physics of phonons*. Taylor Francis Group, 1990.

[50] Y.-L Chen and D.-P. Yang. *Mössbauer effect in lattice dynamics*. Wiley-VCH, 2007.

[51] M. Ortiz and R. Phillips. "Advances in applied mechanics." In: ed. by E. van der Giessen and T.Y. Wu. Vol. 36. Academic Press, 1999. Chap. Nanomechanics of defects in solids.

[52] S.K. Sinha. "Lattice dynamics of copper." In: *Phys. Rev.* 143 (1966), p. 422.

[53] J.L. Warren, J.L. Yarnell, G. Dolling, and R.A. Cowley. "Lattice dynamics of diamond." In: *Phys. Rev.* 158 (1967), p. 805.

[54] M.E. Straumanis and L.S. Yu. "Lattice parameters, densities, expansion coefficients and perfection of structure of Cu and Cu-In alpha phase." In: *Acta Cryst.* A25 (1969), p. 676.

[55] W.C. Overton Jr. and J. Gaffney. "Temperature variation of the elastic constants of cubic elements. I. Copper." In: *Phys. Rev.* 98 (1955), p. 969.

[56] C. Kittel and H. Kroemer. *Thermal physics*. 2nd ed. Freeman, 1980.

[57] F. Reif. *Fundamentals of statistical and thermal physics*. McGraw Hill, 1965.

[58] J.P. Sethna. *Entropy, order parameters, and complexity*. Oxford University Press, 2006.

[59] A. Einstein. "The Planck theory of radiation and the theory of specific heat." In: *Ann. d. Physik* 22 (1906), p. 180.

[60] P. Debye. "The theory of specific warmth." In: *Ann. d. Physik* 39 (1912), p. 789.

[61] D.K. Hsu and R.G. Leisure. "Elastic constants of palladium and beta-phase palladium hydride between 4 and 300-K." In: *Phys. Rev. B* 20 (1979), p. 1339.

[62] M. Moss, P.M. Richards, E.L. Venturini, J.H. Grieske, and E.J. Graeber. "Hydrogen contribution to the heat capacity of single phase, face centered cubic scandium deuteride." In: *J. Chem. Phys.* 84 (1986), p. 956.

[63] L.A. Nygren and R.G. Leisure. "Elastic constants of β-phase PdHx over the temperature range 4–300 K." In: *Phys. Rev. B* 37 (1988), p. 6482.

[64] M.A. Omar. *Elementary solid state physics*. Addison Wesley, 1975.

[65] O.L. Anderson. "A simplified method for calculating Debye temperatures from elastic constants." In: *J. Phys. Chem. Solids* 24 (1963), p. 909.

[66] R.A. Robie and J.L. Edwards. "Some Debye temperatures from single-crystal elastic constant data." In: *J. Appl. Phys.* 37 (1966), p. 2659.

[67] J.J. Adams. "Elastic constants of monocrystal iron form 3 to 500K." MA thesis. Physics Dept., Colorado State Univ., Fort Collins, CO USA, 2006.

[68] J.D. Maynard. "The use of piezoelectric film and ultrasound resonance to determine the complete elastic tensor in one measurement." In: *J. Acoust. Soc. Am.* 91 (1992), p. 1754.

[69] D.I. Bolef and J.C. Miller. "Physical acoustics." In: ed. by W.P. Mason and R.N. Thurston. Vol. VIII. Academic Press, 1971. Chap. 3.

[70] N.F. Foster. "Cadium sulphide evaporated-layer transducers." In: *Proc. IEEE* 53 (1965), p. 1400.

[71] J. de Klerk. "Physical acoustics". In: ed. by W.P. Mason. Vol. IVA. Academic Press, 1966. Chap. 5.

[72] R.G. Leisure and D.I. Bolef. "CW microwave spectrometer for ultrasonic paramagnetic resonance." In: *Rev. Sci. Instrum.* 39 (1968), p. 199.

[73] H.A. Spetzler, G. Chen, S. Whitehead, and I.C. Getting. "A new ultrasonic interferometer for the determination of equation of state parameters of sub-millimeter single crystals." In: *PAGEOPH* 141 (1993), p. 341.

[74] Q. Zhou, S. Lau, D. Wu, and K. Shung. "Piezoelectric films for high frequency ultrasonic transducers in biomedical applications." In: *Prog. Mater. Sci.* 56 (2011), p. 139.

[75] E.P. Pakadakis and T.P. Lerch. "Handbook of elastic properties of solids, liquids and gases." In: ed. by M. Levy, H.E. Bass, and R. Stern. Vol. I. Academic Press, 2001. Chap. 2, p. 39.

[76] S. Eros and J.R. Reitz. "Elastic constants by the ultrasonic pulse echo method." In: *J. Appl. Phys.* 29 (1958), p. 683.

[77] H.J. McSkimin. "Pulse superposition method for measuring ultrasonic wave velocities in solid." In: *J. Acoust. Soc. Am.* 33 (1961), p. 12.

[78] H.J. McSkimin and P. Andreatch. "Analysis of the pulse superposition method for measuring ultrasonic wave velocities as a function of temperature and pressure." In: *J. Acoust. Soc. Am.* 34 (1962), p. 609.

[79] E.P. Papadakis. "Ultrasonic phase velocity by pulse-echo-overlap method incorporating diffraction phase corrections." In: *J. Acoust. Soc. Amer.* 42 (1967), p. 1045.

[80] C. Pantea, D.G. Rickel, A. Migliori, R.G. Leisure, J.Z. Zhang, Y.S. Zhao, S. El-Khatib, and B.S. Li. "Digital ultrasonic pulse-echo overlap system and algorithm for unambiguous determination of pulse transit time." In: *Rev. Sci Instrum.* 76 (2005), p. 114902.

[81] A.E. Petrova and S.M. Stishov. "A digital technique for measuring the velocity and attenuation of sound." In: *Instrum. Exp. Tech.* 52 (2009), p. 609.

[82] H. Niesler and I. Jackson. "Pressure derivatives of elastic wave velocities from ultrasonic interferometric measurements on jacketed polycrystals." In: *J. Acoust. Soc. Am.* 86 (1989), p. 1573.

[83] D.I. Bolef and M. Menes. "Measurement of elastic constants of RbBr, RbI, CsBr, and CsI by an ultrasonic cw resonance technique." In: *J. Appl. Phys.* 31 (1960), p. 1010.

[84] D.I. Bolef. "Elastic constants of single crystals of the bcc transition elements V, Nb, and Ta." In: *J. Appl. Phys.* 32 (1961), p. 100.

[85] D.I. Bolef and J.D. Klerk. "Anomalies in the elastic constants and thermal expansion of chromium single crystals." In: *Phys. Rev.* 129 (1963), p. 1063.

[86] D.I. Bolef and M. Menes. "Nuclear magnetic resonance acoustic absorption in KI and KBr." In: *Phys. Rev.* 114 (1959), p. 1441.

[87] D.I. Bolef. "Physical acoustics." In: ed. by W.P. Mason. Vol. IVA. Academic Press, 1966. Chap. 3.

[88] R.G. Leisure and D.I. Bolef. "Temperarure dependence of ultrasonic paramagnetic resonance in MgO.Fe2+." In: *Phys. Rev. Lett.* 19 (1967), p. 957.

[89] W.C. Cady. *Piezoelectricity*. Vol. Two. Dover, 1964.

[90] A.S. Nowick and B.S. Berry. *Anelastic relaxation in crystalline solids*. Academic Press, 1972.

[91] W. Hermann and H.-G. Sockel. "Handbook of elastic properties of solids, liquids and gases." In: ed. by M. Levy, H.E. Bass, and R.R. Stern. Vol. I. Academic Press, 2001. Chap. 13.

[92] I. Ohno. "Free vibration of a rectangular parallelepiped crystal and its application to determination of elastic constants of orthorhombic crystals." In: *J. Phys. Earth* 24 (1976), p. 355.

[93] A. Migliori, J.L. Sarrao, W.M. Visscher, T.M. Bell, M. Lei, Z. Fisk, and R.G. Leisure. "Resonant ultrasound spectroscopic techniques for measurement of the elastic moduli of solids." In: *Physica B: Conden. Matt.* 183 (1993), p. 1.

[94] A. Migliori and J.L. Sarrao. *Resonant ultrasound spectroscopy*. Wiley, 1997.

[95] A. Migliori and J.D. Maynard. "Implementation of a modern resonant ultrasound spectroscopy system for the measurement of the elastic moduli of small solid specimens." In: *Rev. Sci. Instrum.* 76 (2005), p. 121301.

[96] R.G. Leisure and F.A. Willis. "Resonant ultrasound spectroscopy." In: *J. Condens. Matter* 9 (1997), p. 6001.

[97] H. Ekstein and T. Schiffman. "Free Vibrations of Isotropic Cubes and Nearly Cubic Parallelepipeds." In: *J. Appl. Phys.* 27 (1956), p. 405.

[98] R. Holland and E.P. Eer Nisse. "Variational evaluation of admittances of multielectroded 3-dimensional piezoelectric structures." In: *IEEE Trans. Sonics Ultrasonics* SU-15 (1968), p. 119.

[99] H.H. Demarest, Jr. "Cube resonance method to determine the elastic constants of solids." In: *J. Acoust. Soc. Am.* 49 (1971), p. 768.

[100] P. Heyliger, A. Jilani, H. Ledbetter, R.G. Leisure, and C.-L. Wang. "Elastic constants of isotropic cylinders using resonant ultrasound." In: *J. Acoust. Soc. Am.* 94 (1993), p. 1482.

[101] W.M. Visscher, A. Migliori, T.M. Bell, and R.A. Reinert. "On the normal modes of free vibrations of inhomogeneous and anisotropic elastic objects." In: *J. Acoust. Soc. Am.* 90 (1991), p. 2154.

[102] H. Goldstein. *Classical mechanics*. Addison-Wesley, 1959.

[103] E.P. Eer Nisse. "Resonances of 1-dimensional composite piezoelectric and elastic structures." In: *IEEE Trans. Sonics Ultrasonics* SU-14 (1967), p. 59.

[104] W.H. Press, B.P. Flannery, S.A. Teukolsky, and W.T. Vetterling. *Numerical recipes*. Cambridge University Press, 1986.

[105] https://nationalmaglab.org/user-resources/all-measurement-techniques.

[106] J.R. Taylor. *Classical mechanics*. University Science Books, 2005.

[107] J.B. Mehl. "Analysis of resonance standing-wave measurements." In: *J. Acoust. Soc. Am.* 64 (1978), p. 1523.

[108] R.G. Leisure, K. Foster, J.E. Hightowoer, and D.S. Agosta. "Internal friction studies by resonant ultrasound spectroscopy." In: *Mater. Sci. Eng. A* 370 (2004), p. 34.

[109] C. Thomsen, J. Strait, Z. Vardeny, H.J. Maris, J. Tauc, and J.J. Hauser. "Coherent phonon generation and detection by picosecond light pulses." In: *Phys. Rev. Lett.* 53 (1984), p. 989.

[110] C. Thomsen, H.T. Grahn, H.J. Maris, and J. Tauc. "Surface generation and detection of phonons by picosecond light pulses." In: *Phys. Rev. B* 34 (1986), p. 4129.

[111] H.T. Grahn, H.J. Maris, and J. Tauc. "Picosecond ultrasonics." In: *IEEE J. Quantum Electron.* 25 (1989), p. 2562.

[112] G.L. Eesley, B.M. Clemens, and C.A. Paddock. "Generation and detection of picosecond acoustic pulses in thin metal films." In: *Appl. Phys. Lett.* 50 (1987), p. 717.

[113] O.B. Wright and K. Kawashima. "Coherent phonon detection from ultrafast surface vibrations." In: *Phys. Rev. Lett.* 1668 (1992), p. 2029.

[114] T.C. Zhu, H.J. Maris, and J. Tauc. "Attenuation of longitudinal-acoustic phonons in amorphous SiO_2 at frequencies up to 440 GHz." In: *Phys. Rev. B* 44 (1991), p. 4281.

[115] C. Thomsen, H.T. Grahn, H.J. Maris, and J. Tauc. "Picosecond interferometric technique for study of phonons in the Brillouin frequency range." In: *Optics Commun.* 60 (1986), p. 55.

[116] A. Devos and R. Côte. "Strong oscillations detected by picosecond ultrasonics in silicon: Evidence for an electronic-structure effect." In: *Phys. Rev. B* 70 (2004), p. 125208.

[117] P. Emery and A. Devos. "Strong oscillations detected by picosecond ultrasonics in silicon: Evidence for an electronic-structure effect." In: *Appl. Phys. Lett.* 89 (2006), p. 191904.

[118] A. Devos, M. Foret, S. Ayrinhac, P. Emery, and B. Rufflé. "Hypersound damping in vitreous silica measured by picosecond acoustics." In: *Phys. Rev. B* 77 (2008), 100201.

[119] R. Côte and A Devos. "Refractive index, sound velocity and thickness of thin transparent films from multiple angles picosecond ultrasonics." In: *Rev. Sci. Instrum.* 76 (2005), p. 053906.

[120] H. Ogi, T. Shagawa, N. Nakamura, M. Hirao, H. Okada, and N. Kihara. "Elastic constant and Brillouin oscillations in sputtered vitreous SiO_2 thin films." In: *Phys. Rev. B* 78 (2008), p. 134204.

[121] A. Nagakubo, H. Ogi, H. Ishida, M. Hirao, T. Yokoyama, and T. Nisihara. "Temperature behavior of sound velocity of fluorine-doped vitreous silica thin films studied by picosecond ultrasonics." In: *J. Appl. Phys.* 118 (2015), p. 014307.

[122] T. Dehoux and B. Audoin. "Non-invasive optoacoustic probing of the density and stiffness of single biological cells." In: *J. Appl. Phys.* 112 (2012), p. 124702.

[123] F. Pérez-Coda, R.J. Smith, E. Moradi, L. Marques, K.F. Webb, and M. Clark. "Thin-film optoacoustic transducers for subcellular Brillouin oscillation imaging of individual biological cells." In: *Appl. Opt.* 54 (2015), p. 8388.

[124] K. Shinokita, K. Reimann, M. Woerner, T. Elsaesser, R. Hey, and C. Flytzanis. "Strong amplification of coherent acoustic phonons by intraminiband currents in a semiconductor superlattice." In: *Phys. Rev. Lett.* 116 (2016), p. 075504.

[125] H. Ledbetter. "Materials at low temperatures." In: ed. by R.P. Reed and A.F. Clark. American Society for Metals, 1983. Chap. Elastic properties, pp. 1–45.

[126] A.B. Bhatia. *Ultrasonic absorption*. Clarendon Press, 1967.

[127] C.Y. Ho and R.E. Taylor. *Thermal expansion of solids*. ASM International, 1998.

[128] VASP, The Vienna ab initio simulation package, is available at www.vasp.at/.

[129] Quantum Espresso is available at www.quantum-espresso.org/.

[130] R. Arroyave, D. Shin, and Z.-K. Liu. "Ab initio thermodynamic properties of stoichiometric phases in the NiAl system." In: *Acta Mater.* 53 (2005), p. 1809.

[131] J.A. Garber and A.V. Granato. "Theory of the temperature dependence of second-order elastic constants in cubic materials." In: *Phys. Rev. B* 11 (1975), p. 3990.

[132] G. Steinle-Neumann, L. Stixrude, and R.E. Cohen. "First-principles elastic constants for the hcp transition metals Fe, Co, and Re at high pressure." In: *Phys. Rev. B* 60 (1999), p. 791.

[133] R. Golesorkhtabar, P. Pavone, J. Spitaler, P. Puschnig, and C. Drax. "ElaStic: a tool for calculating second-order elastic constants from first principles." In: *Comput. Phys. Commun.* 184 (2013), p. 1861.

[134] H. Ledbetter. "Sound velocities, elastic constants: temperature dependence." In: *Mater. Sci. Eng., A* 442 (2006), p. 31.

[135] E. I. Isaev, S. I. Simak, A. S. Mikhaylushkin, Yu. Kh. Vekilov, E. Yu. Zarechnaya, L. Dubrovinsky, N. Dubrovinskaia, M. Merlini, M. Hanfland, and I. A. Abrikosov. "Impact of lattice vibrations on equation of state of the hardest boron phase." In: *Phys. Rev. B* 83 (2011), p. 132106.

[136] G. Liebfried and W. Ludwig. "Solid state physics." In: ed. by F. Sietz and D. Turnbull. Vol. 12. Academic Press, 1961, p. 276.

[137] S. Shang, Y. Wang, and Z.K. Liu. "First-principles calculations of phonon and thermodynamic properties in the boron-alkaline earth metal binary systems: B-Ca, B-Sr, and B-Ba." In: *Phys. Rev. B* 75 (2007), p. 024302.

[138] Y. Wang, J. J. Wang, H. Zhang, V. R. Manga, S. L. Shang, L.-Q. Chen, and Z.-K. Liu. "A first-principles approach to finite temperature elastic constants." In: *J. Phys. Condens. Matter* 22 (2010), p. 225404.

[139] K. Kadas, L. Vitos, R. Ahuja, B. Johansson, and J. Kollar. "Temperature dependent elastic properties of alpha beryllium from first principles." In: *Phys. Rev. B* 76 (2007), p. 235109.

[140] W.J. Golumbfskie, R. Arroyave, D. Shin, and Z.-K. Liu. "Finite-temperature thermodynamic and vibrational properties of Al-Ni-Y compounds via first-principles calculations." In: *Acta Mater.* 54 (2006), p. 2291.

[141] L.D. Landau and E.M. Lifshitz. *Statistical physics*. Pergamon Press, 1980.

[142] M. Sob, M. Friak, D. Legut, J. Fiala, and V. Vitek. "The role of ab initio electronic structure calculations in studies of the strength of materials." In: *Mater. Sci. Eng.* A 387 (2004), p. 148.

[143] A. Wang, S.-L. Shang, M. He, Y. Du, L. Chen, R. Zhang, D. Chen, B. Fan, F. Meng, and Z.-K. Liu. "Temperature-dependent elastic stiffness constants of fcc-based metal nitrides from first-principles calculations." In: *J. Mater. Sci.* 49 (2014), p. 424.

[144] D.C. Wallace. "Thermoelasticity of stressed materials and comparison of various elastic constants." In: *Phys. Rev.* 162 (1967), p. 776.

[145] A.F. Jankowski and T. Tsakalakos. "The effect of strain on the elastic constants of noble metals." In: *J.Phys, F: Met. Phys* 15 (1985), p. 1279.

[146] F. Birch. "Finite elastic strain of cubic crystals." In: *Phys. Rev.* 71 (1947), p. 809.

[147] Y. Hiki, J.F. Thomas Jr., and A.V. Granato. "Anharmonicity in noble metals: some thermal properties." In: *Phys. Rev.* 153 (1967), p. 764.

[148] K. Brugger. "Generalized Gruneisen parameters in the anisotroyic Debye model." In: *Phys. Rev.* 137 (1965), A1826.

[149] Z. Wu and R.M. Wentzcovitch. "Quasiharmonic thermal elasticity of crystals: an analytical approach." In: *Phys. Rev. B.* 83 (2011), p. 184115.

[150] D.J. Safarik and R.B. Schwarz. "Evidence for highly anharmonic low-frequency vibrational modes in bulk amorphous $Pd_{40}Cu_{40}P_{20}$." In: *Phys. Rev. B* 80 (2009), p. 094109.

[151] A Migliori, H. Ledbetter, R.G. Leisure, C. Pantea, and J.B. Betts. "Diamonds elastic stiffnesses from 322 K to 10 K." In: *J. Appl. Phys.* 104 (2008), p. 053512.

[152] H.S. Robertson. *Statistical thermophysics*. Prentice Hall, 1993.

[153] B.T. Bernstein. "Electron contribution to the temperature dependence of the elastic constants of cubic metals. I. Normal metals." In: *Phys. Rev.* 132 (1963), p. 50.

[154] G.A. Alers. "Physical acoustics." In: ed. by W.P. Mason. Vol. IV Part A. Academic Press, 1966. Chap. 7.

[155] Y.P. Varshni. "Temperature dependence of the elastic constants." In: *Phys. Rev. B* 2 (1970), p. 3952.

[156] H. Ledbetter. "Relationship between bulk-modulus temperature-dependence and thermal expansivity." In: *Phys. Stat. Solidi B* 181 (1994), p. 81.

[157] P. Toledano and V. Dmitriev. *Reconstructive phase transitions*. World Scientific, 1996.

[158] J.-C. Toledano and P. Toledano. *Landau theory of phase transitions*. World Scientific, 1987.

[159] W. Rehwald. "Study of structural phase-transitions by means of ultrasonic experimets." In: *Advances in Physics* 22 (1973), p. 721,

[160] M.A. Carpenter and E.K.H. Salje. "Elastic anomalies in minerals due to structural phase transitions." In: *Eur. J. Mineral.* 10 (1998), p. 693.

[161] F. Cordero, F. Trequattrini, V. B. Barbeta, R. F. Jardim, and M. S. Torikachvili. "Anelastic spectroscopy study of the metal-insulator transition of $Nd_{1-x}Eu_xNiO_3$." In: *Phys. Rev. B* 84 (2011), p. 125127.

[162] A.V. Granato, K.L. Hultman, and K.-F. Huang. “Ultrasonic response to two and four level quantum systems.” In: *J. Physique* Colloque C10 (1985), pp. C10–C23.

[163] C. Zener. “Stress induced preferential orientation of pairs of solute atoms in metallic solid solution.” In: *Phys. Rev.* 71 (1947), p. 34.

[164] F.M. Mazzolai, P.G. Bordoni, and F.A. Lewis. “Elastic energy-dissipation effects in alpha+beta and beta-phase composition ranges of the palladium-hydrogen system.” In: *J. Phys. F: Met. Phys.* 11 (1981), p. 337.

[165] R.G. Leisure, T. Kanashiro, P.C. Riedi, and D.K. Hsu. “Hydrogen motion in single-crystal palladium hydride as studied ultrasonically.” In: *Phys. Rev. B* 27 (1983), p. 4872.

[166] J.L. Snoek. “Mechanical after effect and chemical constitution.” In: *Physica* 6 (1939), p. 591.

[167] R.G. Leisure, S. Kern, F.R. Drymiotis, H. Ledbetter, A. Migliori, and J.A. Mydosh. “Complete elastic tensor through the first-order transformation in $U_2Rh_3Si_5$.” In: *Phys. Rev. Lett.* 95 (2005), p. 075506.

[168] B. Lüthi, M.E. Mullen, and E. Bucher. “Elastic constants in singlet ground-state Systems: PrSb and Pr.” In: *Phys. Rev. Lett.* 31 (1973), p. 95.

[169] R.S. Lakes. *Viscoelastic solids*. CRC Press, 1998.

[170] M. O’Donnell, E.T. Jaynes, and J.G. Miller. “Kramers-Kronig relationship between ultrasonic-attenuation and phase-velocity.” In: *J. Acoust. Soc. Am.* 69 (1981), p. 696.

[171] V. Mangulis. “Kramers-Kronig or dispersion relations in acoustics.” In: *J. Acoust. Soc. Am.* 36 (1964), p. 211.

[172] V.L. Ginzberg. “Concerning the general relationship between absorption and dispersion of sound waves.” In: *Akust.Zh.* 1 (1955). English translation in *Soviet Phys. Acoust. 1*, 32 (1957), p. 32.

[173] J.P. Wittmer, H. Xu, O. Benzerara, and J. Baschnagel. “Fluctuation-dissipation relation between shear stress relaxation modulus and shear stress autocorrelation function revisited.” In: *Molecular Physics* 113 (2015), p. 2881.

[174] D. Chandler. *Introduction to modern statistical mechanics*. Oxford University Press, 1987.

[175] A.A. Gusev, M.M. Zeldner, and U.W. Suder. “Fluctuation formula for elastic constants.” In: *Phys. Rev. B, Brief Reports* 54 (1996), p. 1.

[176] C. Zener. *Elasicity and anelasicity of metals*. University of Chicago Press, 1948.

[177] M. Callens-Raadschelders, R. De Batist, and R. Gevers. “Debye relaxation equations for a standard linear solid with high relaxation strength.” In: *J. Materials Sci.* 12 (1977), p. 251.

[178] K. Lücke. “Ultrasonic attenuation caused by thermoelastic heat flow.” In: *J. Appl. Phys.* 27 (1956), p. 1433.

[179] B.C. Daly, K. Kang, Y. Yang, and D.G. Cahill. “Picosecond ultrasonic measurements of attenuation of longitudinal acoustic phonons in silicon.” In: *Phys. Rev. B* 80 (2009), p. 174112.

[180] H.E. Bömmel and K. Dransfield. “Excitation and attenuation of hypersonic waves in quartz”. In: *Phys. Rev.* 117 (1960), p. 1245.

[181] J. de Klerk. “Behavior of coherent microwave phonons at low temperatures in Al_2O_3 using vapor-deposited thin-film piezoelectric transducers”. In: *Phys. Rev.* 139 (1965), A 1635.

[182] S. Stoffels, E. Autizi, R. Van Hoof, S. Severi, R. Puers, A. Witvrouw, and H.A.C. Tilmans. “Physical loss mechanisms for resonant acoustical waves in boron doped poly-SiGe deposited with hydrogen dilution.” In: *J. Appl. Phys.* 108 (2010), p. 084517.

[183] L. Landau and G. Rumer. "Uber schallabsorption in festen korpern." In: *Phys. Z. Sowjetunion* 11 (1937), p. 18.
[184] P.G. Klemens. "Physical acoustics." In: ed. by W.P. Mason. Vol. III Part B. Academic Press, 1965. Chap. 5.
[185] H.J. Maris. "Physical acoustics." In: ed. by W.P. Mason and R.N. Thurston. Vol. VIII. Academic Press, 1971. Chap. 6.
[186] A. Akhieser. "On the absorption of sound in solids." In: *J. Phys. (Moscow)* 1 (1939), p. 277.
[187] T.O. Woodruff and H. Ehrenreich. "Absorption of sound in insulators." In: *Phys. Rev.* 123 (1961), p. 1553.
[188] H.E. Bömmel. "Ultrasonic attenuation in superconducting lead." In: *Phys. Rev.* 96 (1954), p. 220.
[189] W.P. Mason. "Ultrasonic attenuation due to lattice-electron interaction in normal conducting metals." In: *Phys. Rev.* 97 (1955), p. 557.
[190] R.W. Morse. "Ultrasonic attenuation in metals by electron relaxation." In: *Phys. Rev.* 97 (1955), p. 1716.
[191] A.B. Pippard. "Ultrasonic attenuation in metals." In: *Phil. Mag.* 46 (1955), p. 1104.
[192] A.B. Pippard. "Theory of ultrasonic attenuation in metals and magnetoacoustic oscillations." In: *Proc. Roy. Soc. A* 257 (1960), p. 165.
[193] J. Bardeen, L.N. Cooper, and J.R. Schrieffer. "Theory of superconductivity." In: *Phys. Rev.* 108 (1957), p. 1175.
[194] M. Gottlieb, M. Garbuny, and C.K. Jones. "Physical acoustics." In: ed. by W. P. Mason and R.N. Thurston. Vol. VII. Academic Press, 1970. Chap. 1.
[195] J.A. Rayne and C.K. Jones. "Physical acoustics." In: ed. by W. P. Mason and R.N. Thurston. Vol. VII. Academic Press, 1970. Chap. 3.
[196] M. Levy. "Physical acoustics." In: ed. by W. P. Mason and R.N. Thurston. Vol. XX. Academic Press, 1992. Chap. 1.
[197] B.W. Roberts. "Physical acoustics." In: ed. by W. P. Mason. Vol. IV Pt. B. Academic Press, 1968. Chap. 10.
[198] L.R. Testardi and J.H. Condon. "Physical acoustics." In: ed. by W. P. Mason and R.N. Thurston. Vol. VIII. Academic Press, 1971. Chap. 2.
[199] Y. Eckstein. "Ultrasonic attenuation in antimony. 1. Geometric resonance." In: *Phys. Rev.* 129 (1963), p. 12.
[200] M.P. Greene, A.R. Hoffman, A. Houghton, and J.J. Quinn. "Ultrasonic attenuation in oblique magnetic fields." In: *Phys. Rev.* 156 (1967), p. 798.
[201] M.H. Cohen, M.J. Harrison, and W.A. Harrison. "Magnetic-field dependence of the ultrasonic attentuation in metals." In: *Phys. Rev.* 117 (1960), p. 937.
[202] J.R. Hook and H.E. Hall. *Solid state physics*. Wiley, 1991.
[203] H.P. Myers. *Introductory solid state physics*. Taylor and Francis, 1991.
[204] M. Mongy. "Quantum oscillations of ultrasonic attenuation in gold." In: *J. Phys. Chem. Solids* 33 (1972), p. 1355.
[205] W.D. Wallace and H.V. Bohm. "Quantum oscillations in attenuation of transverse ultrasonic waves in field-cooled chromium." In: *J. Phys. Chem. Solids* 29 (1968), p. 721.
[206] A. Granato and K. Lücke. "Theory of mechanical damping due to dislocations." In: *J. Appl. Phys.* 27 (1956), p. 583.
[207] T.A. Read. "Internal friction of single crystals of copper and zinc." In: *Trans. Am. Inst. Mining Met. Engrs.* 143 (1941), p. 30.
[208] J.S. Koehler. *Imperfections in nearly perfect crystals*. Wiley, 1952.

[209] A.S. Nowick. "Internal friction and dynamic modulus of cold-worked metals." In: *J. Appl. Phys.* 25 (1954), p. 1129.

[210] A.V. Granato and K. Lücke. "Physical acoustics." In: ed. by W.P. Mason. Vol. IV Part A. Academic Press, 1966. Chap. 6.

[211] R.M. Stern and A.V. Granato. "Overdamped resonance of dislocations in copper." In: *Acta Met.* 10 (1962), p. 358.

[212] C. Wert and C. Zener. "Interstitial atomic difusion coefficients." In: *Phys. Rev.* 76 (1949), p. 1169.

[213] C.R. Ko, K. Salama, and J.M. Roberts. "Effect of hydrogen on the temperature-dependence of the elastic constants of vanadium single crystals." In: *J. Appl. Phys.* 51 (1980), p. 1014.

[214] J. Buchholz, J. Völkl, and G. Alefeld. "Anomalously small elastic Curie constant of hydrogen in tantulam." In: *Phys. Rev. Lett.* 30 (1973), p. 318.

[215] K. Foster, J.E. Hightower, R.G. Leisure, and A.V. Skripov. "Ultrasonic attenuation and dispersion due to hydrogen motion in the C15 Laves-phase compound TaV_2H_x." In: *J. Phys.:Condensed Matter* 13 (2001), p. 7327.

[216] R.G. Leisure, T. Kanashiro, P.C. Riedi, and D.K. Hsu. "An ultrasonic study of stress-induced ordering of hydrogen in single-crystal palladium hydride." In: *J. Phys. F: Met. Phys.* 13 (1983), p. 2025.

[217] R.G. Leisure, L.A. Nygren, and D.K. Hsu. "Ultrasonic relaxation rates in palladium hydride and palladium deuteride." In: *Phys. Rev. B* 33 (1986), p. 8325.

[218] J.O. Fossum and K. Fossum. "Measurements of ultrasonic attenuation and velocity in Verneuil-grown and flux grown $SrTiO_3$." in: *J. Phys. C: Solid State Phys.* 18 (1985), p. 5549.

[219] L.D. Landau and I.M. Khalatnikov. In: *Sov. Phys. Dokl.* 96 (1954), p. 469.

[220] J.O. Fossum. "A phenomenological analysis of ultrasound near phase transitions." In: *J. Phys. C: Solid State Phys.* 18 (1985), p. 5531.

[221] A. Abragam. *The principles of nuclear magnetism.* Oxford University Press, 1961.

[222] C.P. Slichter. *The principles of magnetic resonancee.* Harper and Row, 1963.

[223] R.A. Alpher and R.J. Rubin. "Magnetic dispersion and attenuation of sound in conducting fluids and solids." In: *J. Acoust. Soc. Am.* 26 (1954), p. 452.

[224] J. Buttet, E.H. Gregory, and P.K. Bailey. "Nuclear acoustic resonance in aluminum via coupling to magnetic dipole moment." In: *Phys. Rev. Lett.* 23 (1969), p. 1030.

[225] R.G. Leisure, D.K. Hsu, and B.A. Seiber. "Nuclear-acoustic-resonance absorption and dispersion in aluminum." In: *Phys. Rev. Lett.* 30 (1973), p. 1326.

[226] R. Guermeur, J. Joffrin, A. Levelut, and J. Penne. "Mesure de la variation de la vitesse de phase dune onde ultrasonore se propageant dans un cristal paramagnetique." In: *Phys. Lett.* 13 (1964), p. 107.

[227] E.B. Tucker. "Physical acoustics." In: ed. by W.P. Mason. Vol. IV Pt. A. Academic Press, 1966. Chap. 2.

[228] R.C. Zeller and R.O. Pohl. "Thermal conductivity and specific heat of noncrystalline solids." In: *Phys. Rev. B* 4 (1971), p. 2029.

[229] P.W. Anderson, B.I. Halperin, and C.M. Varma. "Anomalous low-temperature thermal properties of glasses and spin glasses." In: *Phil. Mag.* 25 (1972), p. 1.

[230] W.A. Phillips. "Tunneling states in amorphous solids." In: *J. Low Temp. Phys.* 7 (1972), p. 351.

[231] J. Jäckle. "Ultrasonic attenuation in glasses at low temperatures." In: *Z. Physik* 257 (1972), p. 212.

[232] P. Doussineau, C. Frénois, R.G. Leisure, A. Levelut, and J.-Y. Prieur. "Amorphous-like acoustical properties of Na doped beta Al_2O_3." In: *J. Phys (Paris)* 41 (1980), p. 1193.

[233] S.N. Coppersmith and B. Golding. "Low-temperature acoustic properties of metallic glasses." In: *Phys. Rev. B* 47 (1993), p. 4922.

[234] A.J. Leggett and D.C. Vural. "Tunneling two-level systems model of low-temperature properties of glasses: Are smoking-gun tests possible?" In: *J. Phys. Chem. B* 117 (2013), p. 12966.

[235] D.R. Queen, X. Liu, J. Karel, T.H. Metcalf, and F. Hellman. "Excess specific heat in evaporated amorphous silicon." In: *Phys. Rev. Lett.* 110 (2013), p. 135901.

[236] G.S. Kino. *Acoustic waves: devices, imaging and analog signal processing*. Prentice-Hall, 1987.

[237] T.F. Hueter and R.H. Bolt. *Sonics*. Wiley, 1955.

[238] C.G. Montgomery, R.H. Dicke, and E.M. Purcell. *Principles of microwave circuits*. The Institute of Engineering and Technology, 1987, p. 67.

[239] E.P. Papadakis. "Ultrasonic diffraction loss and phase change in anisoptric materials." In: *J. Acoust. Soc. Am.* 40 (1966), p. 863.

[240] E.P. Papadakis. "Physical acoustics." In: ed. by W. P. Mason and R.N. Thurston. Vol. XI. Academic Press, 1975. Chap. 3.

[241] P.R. Saulson. "Thermal noise in mechanical experiments." In: *Phys. Rev. D* 42 (1990), p. 2437.

[242] J.R. Neighbours and G.A. Alers. "Elastic constants of silver and gold." In: *Phys. Rev.* 111 (1958), p. 707.

[243] Y.S. Touloukian, R.K. Kirby, R.E. Taylor, and P.D. Desa. *Thermophysical properies of matter*. Vol. 12. Plenum, 1975.

[244] Y. Hiki and A.V. Granato. "Anharmonicity in noble metals; higher order elastic constants." In: *Phys. Rev.* 144 (1966), p. 411.

[245] J.M. Lang, Jr. and Y.M. Gupta. "Experimental determination of third-order elastic constants of diamond." In: *Phys. Rev. Lett.* 106 (2011), p. 125502.

[246] C. Giles, C. Adriano, A.F. Lubambo C. Cusatis I. Mazzaro and M.G. Honnicke. "Diamond thermal expansion measurement using transmitted X-ray back-diffraction." In: *J. Synchrotron Radiation* 12 (2005), p. 349.

[247] J. Philip and M.A. Breazeale. "Third order elastic constants and Grüneisen parameters of silicon and germanium between 3 and 300K." In: *J. Appl. Phys.* 54 (1983), p. 752.

[248] K. Salama and G.A. Alers. "Third-order elastic constants of copper at low temperature." In: *Phys. Rev.* 161 (1967), p. 673.

Index